OCR

RECOGNISING ACHIEVEMENT

G C S E
Mathematics

Graduated Assessment

Stages 1 & 2

Authors

Howard Baxter

Mike Handbury

John Jeskins

Jean Matthews

Mark Patmore

Contributor

Colin White

Series editor *Brian Seager*

Hodder & Stoughton

A MEMBER OF THE HODDER HEADLINE GROUP

Orders: please contact Bookpoint Ltd, 130 Milton Park, Abingdon, Oxon OX14 4SB.
Telephone: (44) 01235 827720, Fax: (44) 01235 400454. Lines are open from
9.00 – 6.00, Monday to Saturday, with a 24 hour message answering service.
Email address: orders@bookpoint.co.uk

British Library Cataloguing in Publication Data

A catalogue record for this title is available from The British Library.

ISBN 0 340 801883

First published 2001

Impression number 10 9 8 7 6

Year 2007 2006 2005 2004

Cover illustration by Mike Stones.

Produced by Hardlines, Charlbury, Oxon.

Printed in Dubai for Hodder & Stoughton Educational, a division of Hodder
Headline Plc, 338 Euston Road, London NW1 3BH.

This book covers the first part of the specification for the Foundation tier of GCSE Mathematics. It is particularly aimed at OCR Mathematics C (Graduated Assessment) but could be used for other GCSE Mathematics examinations.

The work in this book covers the criteria in stages M1 and M2, and aims to make the best of your performance in the module tests and the terminal examination:

- Each chapter is presented in a style intended to help you understand the mathematics, with straightforward explanations and worked examples.
- At the start of most chapters is a list of what you should already know before you begin.
- There are plenty of exercises for you to work through and practise the skills.
- At the end of each chapter there is a list of key ideas.
- After every two or three chapters there is a revision exercise.
- Some exercises are designed to be done without a calculator so that you can practise for the non-calculator sections of the papers.
- Many chapters contain Activities to help you develop the necessary skills to undertake coursework.
- At frequent intervals throughout the book there are exam tips, where the experienced examiners who have written this book offer advice and tips to improve your examination performance.
- Revision exercises and Module tests are provided in the Teacher's Resource.

Part of the examination is a calculator-free zone. You will have to do the first section of each paper without a calculator and the questions are designed appropriately.

Some questions will be set which offer you little help to get started. These are called 'unstructured' or 'multi-step' questions. Instead of the question having several parts, each of which leads to the next, you have to work out the necessary steps to find the answer. There will be examples of this kind of question in the revision tests and past examination papers.

Top ten tips

Here are some general tips from the examiners to help you do well in your tests and examination.

Practise:

1 **taking time** to work through each question carefully
2 answering questions **without** a calculator
3 answering questions which require **explanations**
4 answering **unstructured** questions
5 **accurate** drawing and construction
6 answering questions which **need a calculator**, trying to use it efficiently
7 **checking answers**, especially for reasonable size and degree of accuracy
8 making your work **concise** and well laid out
9 checking that you have **answered the question**
10 **rounding** numbers, but only at the appropriate stage.

Coursework

The GCSE Mathematics examinations will assess your ability to use your mathematics on longer problems than those normally found on timed written examination papers. Assessment of this type of work will account for 20% of your final mark. It will involve two tasks, each taking about three hours. One task will be an investigation, the other a statistics task.

Each type of task has its own mark scheme in which marks are awarded in three categories or 'strands'. The titles of these strands give you clues about the important aspects of this work.

For the investigation tasks the strands are:

- Making and monitoring decisions – what you are going to do and how you will do it
- Communicating mathematically – explaining and showing exactly what you have done
- Developing the skills of mathematical reasoning – using mathematics to analyse and prove your results.

The table below gives some idea of what you will have to do and show. Look at this table whenever you are doing some extended work and try to include what it suggests you do.

Mark	Making and monitoring decisions	Communicating mathematically	Developing the skills of mathematical reasoning
1	organising work, producing information and checking results	discussing work using symbols and diagrams	finding examples that match a general statement
2	beginning to plan work, choosing your methods	giving reasons for choice of presentation of results and information	searching for a pattern using at least three results
3	finding out necessary information and checking it	showing understanding of the task by using words, symbols, diagrams	explaining reasoning and making a statement about the results found

Mark	Making and monitoring decisions	Communicating mathematically	Developing the skills of mathematical reasoning
4	simplifying the task by breaking it down into smaller stages	explaining what the words, symbols and diagrams show	testing generalisations by checking further cases
5	introducing new questions leading to a fuller solution	justifying the means of presentation	justifying solutions explaining why the results occur
6	using a range of techniques and reflecting on lines of enquiry and methods used	using symbolisation consistently	explaining generalisations and making further progress with the task
7	analysing lines of approach and giving detailed reasons for choices	using symbols and language to produce a convincing and reasoned argument	report includes mathematical justifications and explanations of the solutions to the problem
8	exploring extensively an unfamiliar context or area of mathematics and applying a range of appropriate mathematical techniques to solve a complex task	using mathematical language and symbols efficiently in presenting a concise reasoned argument	providing a mathematically rigorous justification or proof of the solution considering the conditions under which it remains valid

For the statistical tasks the strands are:

- Specifying the problem and planning – choosing or defining a problem and outlining the approach to be followed
- Collecting, processing and representing data – explaining and showing what you have done
- Interpreting and discussing results – using mathematical and statistical knowledge and techniques to analyse, evaluate and interpret your results and findings.

The marks obtained from each task are added together to give a total out of 48.

The table below gives some idea of what you will have to do and show. Look at this table whenever you are doing some extended work and try to include what it suggests you do.

Mark	Specifying the problem and planning	Collecting, processing and representing data	Interpreting and discussing results
1–2	choosing a simple problem and outlining a plan	collecting some data; presenting information, calculations and results	making comments on the data and results
3–4	choosing a problem which allows you to use simple statistics and plan the collection of data	collecting data and then processing it using appropriate calculations involving appropriate techniques; explaining what the words, symbols and diagrams show	explaining and interpreting the graphs and calculations and any patterns in the data

Mark	Specifying the problem and planning	Collecting, processing and representing data	Interpreting and discussing results
5–6	considering a more complex problem and using a range of techniques and reflecting on the method used	collecting data in a form that ensures they can be used; explaining statistical meaning through the consistent use of accurate statistics and giving a reason for the choice of presentation; explaining features selected	commenting on, justifying and explaining results and calculations; commenting on the methods used
7–8	analysing the approach and giving reasons for the methods used; using a range of appropriate statistical techniques to solve the problem	using language and statistical concepts effectively in presenting a convincing reasoned argument; using an appropriate range of diagrams to summarise the data and show how variables are related	correctly summarising and interpreting graphs and calculations and making correct and detailed inferences from the data; appreciating the significance of results obtained and, where relevant, allowing for the nature and size of the sample and any possible bias; evaluating the effectiveness of the overall strategy and recognising limitations of the work done, making suggestions for improvement

Advice

Starting a task

Ask yourself:

- what does the task tell me?
- what does it ask me?
- what can I do to get started?
- what equipment and materials do I need?

Working on the task

- Make sure you explain your method and present your results as clearly as possible
- Break the task down into stages. For example in 'How many squares on a chessboard', begin by looking at 1 × 1 squares then 2 × 2 squares, then 3 × 3 squares. In a task asking for the design of a container, start with cuboids then nets, surface area, prisms … Or in statistics you might want to start with a pilot survey or questionnaire.
- Write down questions that occur to you, for example, *what happens if you change the size of a rectangle systematically?* They may help you find out more about the work. In a statistical task you might wish to include different age groups or widen the type of data.

- Explore as many aspects of the task as possible.
- Develop the task into new situations and explore these thoroughly.
  - What connections are possible?
  - Is there a result to help me?
  - Is there a pattern?
  - Can the problem be changed? If so, how?

Explain your work

- Use appropriate words and suitable tables, diagrams, graphs, calculations.
- Link as much of your work together as possible, explaining, for example, why you chose the tables and charts you used and rejected others, or why the median is more appropriate than the mean in a particular statistical analysis, or why a pie chart is not appropriate. Don't just include diagrams to show identical information in different ways.
- Use algebra or symbols to give clear and efficient explanations; in investigations, you must use algebra to progress beyond about 4 marks. You will get more credit for writing $T = 5N + 1$ than for writing 'the total is five times the pattern number, plus one'.
- Don't waffle or use irrelevant mathematics; present results and conclusions clearly.

State your findings

- Show how patterns have been used and test conclusions.
- State general results in words and explain what they mean.
- Write formulae and explain how they have been found from the situations explored.
- Prove the results using efficient mathematical methods.
- Develop new results from previous work and use clear reasoning to prove conclusions.
- Make sure your reasoning is accurate and draws upon the evidence you've presented.
- Show findings in clear, relevant diagrams.
- Check you've answered the question or hypothesis.

Review/conclusion/extension

- Is the solution acceptable?
- Can the task be extended?
- What can be learned from it?

Example task

On the next page there is a short investigative task for you to try, in both 'structured' and 'unstructured' form. The structured form shows the style of a question that might appear on a timed written paper. The unstructured form represents the usual style of a coursework task. The structured form leads you to an algebraic conclusion. Notice the appearance of algebra from question 4 onwards, through a series of structured questions. These mirror the sort of questions you would be expected to think of (and answer) if you were trying it as coursework.

Comments about the questions, linking the two forms of presentation, are also shown.

Although the task in both forms directs you to investigate trapezium numbers, you would be expected to extend the investigation into other forms of number, such as pentagon numbers, to achieve the higher marks.

ACTIVITY

structured form

Trapezium numbers

These diagrams represent the first three trapezium numbers.

Each diagram always starts with two dots on the top row.

So the third trapezium number is 9 because nine dots can be arranged as a trapezium. There are two dots in the top row, three dots in the next row and four dots in the bottom row.

1   Write down the next two trapezium numbers

2   **a)** Draw a table, graph or chart of all the trapezium numbers, from the first to the tenth.
   b) Work out the eleventh trapezium number.

3   The 19th trapezium number is 209. Explain how you could work out the 20th trapezium number without drawing any diagrams.

4   Find an expression for the number of dots in the bottom row of the nth trapezium number. Test your expression for a suitable value of n.

5   Find, giving an explanation, an expression for the number of dots in the bottom row of the diagram for the $(n + 1)$th trapezium number.

6   The nth trapezium number is x. Write down an expression in terms of x and n for the $(n + 1)$th trapezium number. Test your expression for a suitable value of n.

unstructured form

Trapezium numbers

These diagrams represent the first three trapezium numbers.

Each diagram starts with two dots on the top row.

So the third trapezium number is 9 because nine dots can be arranged as a trapezium.

Investigate trapezium numbers

NB Although the task in this form asks you to investigate trapezium numbers, you have the freedom to – and are expected to – extend the investigation to consider other forms of number such as pentagon numbers.

Commentary

This question allows you to show understanding of the task, systematically obtaining information which **could** enable you to find an expression for trapezium numbers.

This question provides a structure, using symbols, words and diagrams, from which you should be able to derive an expression from either a table or a graph. Part **b)** could be done as a 'predict and test'.

In the unstructured form you would not normally answer a question like this.

From here you are **directed** in the structured task, and **expected** in the unstructured task, to use algebra, testing the expression – the **generalisation**.

In the unstructured form this would represent the sort of 'new' question you might ask, to lead to a further solution and to demonstrate symbolic presentation and the ability to relate the work to the physical structure, rather than doing all the analysis from a table of values.

Stage 1

Chapter 1   **Numbers**   **1**
              Whole numbers   1
              Place value   1
              Large numbers   4
              Rounding to the nearest 10, 100, 1000, etc.   6
              Odd and even numbers   8
              5s and 10s   9
              Simple number facts   10

Chapter 2   **Probability**   **18**
              Chance   18

Revision exercise A1   **25**

Chapter 3   **Direction and position**   **27**
              Graphs, coordinates and direction   30

Chapter 4   **Scales**   **39**
              Telling the time   39
              How long does it take?   42
              Reading scales   44

Revision exercise B1   **51**

CONTENTS

Chapter 5	**Algebra patterns and using letters for numbers**	**53**
	Letters for numbers	53
	Number patterns	66
	Function machines	70
Chapter 6	**Solving problems**	**79**
	Fractions	81
Revision exercise C1		**88**
Chapter 7	**Shapes**	**90**
	Triangles	90
	Quadrilaterals	91
	Polygons	92
	Enlargements	95
Chapter 8	**Units, perimeter, area and volume**	**101**
	Lengths	101
	Perimeter, area and volume	107
Revision exercise D1		**122**
Chapter 9	**Representing data**	**124**
	Pictograms	124
	Reading from graphs	128
Chapter 10	**Listing**	**131**
Revision exercise E1		**133**
Stage 2		**137**

1 Numbers

Whole numbers

There are 26 letters in the English alphabet.

Letters are used to make words.

You can change the order to make a different word.

Changing the order of the letters of the word TRAP gives different words, such as PART and RAPT.

To make numbers there are 10 **digits**:

0   1   2   3   4   5   6   7   8   9

Digits are used to make numbers.

As with letters, you can change the order to make a different number.

Place value

> **Exam tip**
>
> To write a number in words, always write *and* after the word hundred.

EXAMPLE 1

How many different numbers can be made using all of the digits 1, 2 and 3? Write each number in digits and in words.

Which is the largest and which is the smallest of these numbers?

Hundreds	Tens	Units	In words
1	2	3	one hundred and twenty-three
1	3	2	one hundred and thirty-two
2	1	3	two hundred and thirteen
2	3	1	two hundred and thirty-one
3	1	2	three hundred and twelve
3	2	1	three hundred and twenty-one

The largest is 321. The smallest is 123.

ACTIVITY 1

- Can you find the largest and smallest numbers without writing them all down? How did you do it?
- How do you know that you have written down all the possible numbers?

Exam tip

To find the largest number, two of the numbers start with 3 in the hundreds position, so the larger number in the tens position will tell you which is largest. To find the smallest number, two of the numbers start with 1 in the hundreds position, so the smaller number in the tens position will tell you which is the smallest.

EXERCISE 1.1A

1   How many different numbers can you make using the following digits?
    Write each number in words alongside each set of digits.
    Write down the largest and the smallest numbers made.
    a) 2, 8, 7   **b)** 4, 1, 9   **c)** 5, 5, 8
2   Write the following numbers using digits.
    a) three hundred and seventy-five
    b) six hundred and ten
    c) four hundred and seven
    d) eight hundred and fourteen
    e) one hundred and thirty
3   Write the following numbers in words.
    a) 56   **b)** 281   **c)** 109
    d) 450   **e)** 700
4   Write down the value of the digit shown in bold.
    a) 4**3**6   **b)** **7**5   **c)** **2**58
    d) 57**9**   **e)** **4**21

EXERCISE 1.1B

1 How many different numbers can you make using the following digits?
Write each number in words alongside each set of digits.
Write down the largest and the smallest numbers made.
   a) 6, 5, 3   **b)** 8, 9, 7   **c)** 4, 2, 0

2 Write the following numbers using digits.
   a) one hundred and forty-two
   b) three hundred and seventy-three
   c) two hundred and twelve
   d) six hundred and forty
   e) eight hundred and nine

3 Write the following numbers in words.
   a) 436   **b)** 851   **c)** 715   **d)** 304   **e)** 92

4 Write down the value of the digit shown in bold.
   a) 552   **b)** 7**9**2   **c)** 30**6**   **d)** 524   **e) 2**90

Large numbers

Large numbers like 8 765 432 are usually written in groups of three digits with a small space between each group. A place value table can help to make sense of the number.

Millions	Hundred thousands	Ten thousands	Thousands	Hundreds	Tens	Units
8	7	6	5	4	3	2

This tells you the number is:

'eight million, seven hundred and sixty-five thousand, four hundred and thirty-two'.

ACTIVITY 2

- Choose any two digits
- Make the largest and the smallest number using these digits. You cannot repeat a digit in either of your numbers.
- Subtract the smallest number from the largest number.
- Repeat this for other sets of two digits.
- What do you notice about your answers?

Exam tip

Notice how the commas in the writing match with the spaces in the number.

ACTIVITY – EXTENSION

Repeat this for numbers made from three digits.

EXERCISE 1.2A

1 Write these numbers in words.
  a) 2645
  b) 98 352
  c) 4 398 526
  d) 55 705
  e) 130 016

2 Write these numbers using digits.
  a) one million, three hundred and seventy-eight thousand, two hundred and fifty-five
  b) four hundred and thirty-six thousand, eight hundred and fifty-one
  c) three million, six hundred and fifty-six thousand, nine hundred and twelve
  d) four million, five hundred thousand
  e) two hundred and seventy-five thousand and fifty

EXERCISE 1.2B

1 Write these numbers in words.
  a) 45 540
  b) 2821
  c) 107 091
  d) 1 486 151
  e) 80 904

2 Write these numbers using digits.
  a) two million, five hundred and ninety-eight thousand, seven hundred and seventy-two
  b) seven hundred and fourteen thousand, five hundred and forty-nine
  c) five million, three hundred and forty-one thousand, eight hundred and thirteen
  d) six hundred and forty thousand, and five
  e) seven hundred thousand, seven hundred

Rounding to the nearest 10, 100, 1000, etc.

Often numbers we see are not exact values. They have been rounded.

This newspaper headline does not mean that exactly 25 000 fans saw the match.

It was probably slightly more or slightly less than this.

> **25 000 FANS SEE UNITED CUP WIN**
>
> It was a capacity crowd that cheered United on Saturday, to a 3-1 win in this year's first round FA Cup competition.

Giving an exact value, such as 24 891, would not make such a dramatic headline.

EXAMPLE 2

Round 2376   **a)** to the nearest 10   **b)** to the nearest 100   **c)** to the nearest 1000

a) To the nearest 10: 2376 is between 2370 and 2380.

```
        2370            2376            2380
```

2376 is nearer to 2380.

b) To the nearest 100: 2376 is between 2300 and 2400.

```
        2300          2376      2400
```

2376 is nearer to 2400.

c) To the nearest 1000: 2376 is between 2000 and 3000.

```
        2000    2376              3000
```

2376 is nearer to 2000.

Exam tip

A number exactly halfway is always **rounded UP**. So 2500 to the nearest 1000 would be 3000; 2350 to the nearest 100 would be 2400.

EXERCISE 1.3A

1   Write these newspaper headlines using a suitable approximate number. Say what accuracy you have used in each case.
   a) 41 638 see Rovers win!
   b) 4327 at pop concert!
   c) MP elected with a 14 873 majority!
   d) Exports exceed £4 683 421!
   e) Bank makes £12 321 457 profit!
   f) Local man wins £998 321 on the lottery!

2   Round each of the following numbers to the nearest million.
   a) 1 804 673
   b) 6 299 888
   c) 3 500 000
   d) 5 200 451
   e) 3 300 033

EXERCISE 1.3B

1   Round each of the following numbers
   (i) to the nearest 10
   (ii) to the nearest 100
   (iii) to the nearest 1000.
   a) 1066   **b)** 23 629   **c)** 8912
   d) 26 788   **e)** 46 950

2   Round each of the following numbers to the nearest million.
   a) 1 500 070   **b)** 5 020 469   **c)** 2 964 720
   d) 4 199 689   **e)** 876 543

3   5 637 812 can be written as
   a) 6 000 000
   b) 5 600 000
   c) 5 637 800
   What is the accuracy used in each of these values?

Odd and even numbers

The counting numbers are made up of two sequences of numbers

1		3		5		7		9		11		13

	2		4		6		8		10		12

Each of these sequences goes up by 2.

> 1, 3, 5, 7, 9, 11, 13, ... are called **odd numbers** and
>
> 2, 4, 6, 8, 10, 12, ... are called **even numbers**.

You may have noticed how all even numbers end in 2, 4, 6, 8 or 0 and
all odd numbers end in 1, 3, 5, 7 or 9.

So   429 is an odd number since it ends in 9 and
      150 is an even number since it ends in 0.

EXERCISE 1.4A

1 List all odd numbers between 20 and 40.
2 List all even numbers between 36 and 53.
3 Make lists of the odd and the even numbers that are in the 3 times table.
4 Make lists of the odd and the even numbers that are in the 4 times table.
5 Is there a times table where all the answers are odd?

EXERCISE 1.4B

1 List all odd numbers between 55 and 70.
2 List all even numbers between 26 and 41.
3 Make lists of the odd and the even numbers that are in the 6 times table.
4 Make lists of the odd and the even numbers that are in the 5 times table.
5 Is there a times table where all the answers are even?

ACTIVITY 3

1 Add two even numbers together.
   Do this several times with different numbers.
   What do you notice about your answers?

2 Add two odd numbers together.
   Do this several times with different numbers.
   What do you notice about your answers?

3 Add one odd and one even number together.
   Do this several times with different odd and even numbers.
   What do you notice about your answers?

5s and 10s

The 5 times table goes	5, 10, 15, 20, 25, 30, 35, 40, ...	and
the 10 times table goes	10, 20, 30, 40, 50, 60, 70, 80, ...	

Notice how the numbers in the 5 times table all end in 0 or 5 and the numbers in the 10 times table all end in 0.

We say that the numbers in the 5 times table are divisible by 5 and the numbers in the 10 times table are divisible by 10.

EXERCISE 1.5A

1 Look at the following list of numbers
   55   67   150   38   400   125
   Write down:
   a) those numbers that are divisible by 5
   b) those numbers that are divisible by 10.

2 From the following list of numbers, write down:
   a) the numbers which are odd
   b) the numbers which are even
   c) the numbers which are divisible by 5
   d) the numbers which are divisible by 10.
   52   305   210   47   16   100   78   35

EXERCISE 1.5B

1 Look at the following list of numbers

    60   245   2000   130   50   105

Write down:

a) those numbers that are divisible by 5

b) those numbers that are divisible by 10.

2 From the following list of numbers, write down:

a) the numbers which are odd.

b) the numbers which are even

c) the numbers which are divisible by 5

d) the numbers which are divisible by 10.

    625   14   30   58   115   710   1000   503

Simple number facts

Addition and subtraction methods

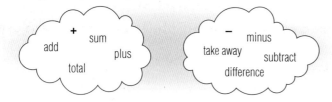

There are various 'mental' ways to do simple addition and subtraction. You may have seen these before.

EXAMPLE 3

Work out   **a)** 23 + 36

              b) 58 + 34

a) Since 36 is 30 + 6, add the two parts on separately.

    23 + 30 = 53   53 + 6 = 59

b) Similarly

    58 + 30 = 88   88 + 4 = 92

EXERCISE 1.6A

Work out

1   26 + 47
2   38 + 53
3   49 + 24
4   63 + 29
5   46 + 36

6   Add 16 and 22.
7   What is 39 plus 24?
8   What is the sum of 69 and 31?
9   Find the total of 57 and 42.
10 Add together 24 and 55.

EXERCISE 1.6B

Work out

1   24 + 31
2   52 + 36
3   26 + 69
4   43 + 28
5   86 + 26

6   Add 61 and 25.
7   What is 74 plus 68?
8   What is the sum of 54 and 17?
9   Find the total of 39 and 24.
10 Add together 74 and 22.

EXAMPLE 4

Work out 55 − 28.
Since 28 is 20 + 8, first take away 20, then take away 8.
55 − 20 = 35   35 − 8 = 27

EXERCISE 1.7A

Work out

1 39 – 23
2 86 – 21
3 48 – 27
4 73 – 54
5 55 – 39

6 What is 59 take away 26?
7 Find 86 minus 69.
8 What is the difference between 80 and 13?
9 What is 93 subtract 45?
10 Work out 72 take away 48.

EXERCISE 1.7B

Work out

1 57 – 14
2 38 – 25
3 74 – 53
4 83 – 39
5 72 – 19
6 What is 46 take away 28?

7 Find 68 minus 43.
8 What is the difference between 94 and 73?
9 What is 83 subtract 78?
10 Work out 70 take away 37.

EXAMPLE 5

In Example 4

55 – 28 = 27

Notice also that 55 – 27 = 28

and that 27 + 28 = 55

The three numbers are linked together and can make three different calculations.

Knowing one tells you the answer to two others.

EXERCISE 1.8A

Use the following numbers to make three different calculations.

1 46   28   18
2 79   91   12
3 12   24   36
4 18   37   19
5 100   28   72

EXERCISE 1.8B

Use the following numbers to make three different calculations.

1 41   56   15
2 56   19   75
3 17   84   67
4 43   62   19
5 100   19   81

EXAMPLE 6

How many do you need to add to

   a) 61 to make 100?

   b) 24 to make 55?

a) First add to get to the next ten ($+9$). Then add tens to get to 100 ($+30$).

  $9 + 30 = 39$

  or   Find $100 - 61$.

  $100 - 60 = 40$ $40 - 1 = 39$

b) First $+6$ (to make up to 30). Then $+20$ (to make up to 50). Then $+5$ (to get 55).

  $6 + 20 + 5 = 31$

  or   Find $55 - 24$.

  $55 - 20 = 35$ $35 - 4 = 31$

EXERCISE 1.9A

How many do you need to add to

1 34 to make 100
2 58 to make 100
3 17 to make 100
4 23 to make 65
5 49 to make 98?

EXERCISE 1.9B

How many do you need to add to

1 61 to make 100
2 77 to make 100
3 18 to make 100
4 24 to make 73
5 51 to make 93?

Multiplication and division methods

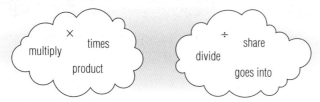

Finding the answer to any multiplication or division sum depends upon knowing your times tables.

If you still do not know all your tables off by heart, now is the time to learn them.

You can use this multiplication grid to help. Cover it up and try to answer the questions without it.

x	1	2	3	4	5	6	7	8	9	10
1	1	2	3	4	5	6	7	8	9	10
2	2	4	6	8	10	12	14	16	18	20
3	3	6	9	12	15	18	21	24	27	30
4	4	8	12	16	20	24	28	32	36	40
5	5	10	15	29	25	30	35	40	45	50
6	6	12	18	24	30	36	42	48	54	60
7	7	14	21	28	35	42	49	56	63	70
8	8	16	24	32	40	48	56	64	72	80
9	9	18	27	36	45	54	63	72	81	90
10	10	20	30	40	50	60	70	80	90	100

Look at the line from the top left-hand corner to the bottom right-hand corner. The numbers are the same on both sides of the line. Can you see why? The order in which the numbers are multiplied does not change the answer.

$8 \times 7 = 56$ and $7 \times 8 = 56$.

EXERCISE 1.10A

Copy these into your book and write down the answers.

1	3 × 4	**8**	5 × 3
2	6 × 2	**9**	7 × 4
3	9 × 5	**10**	8 × 5
4	8 × 6	**11**	6 × 2
5	4 × 7	**12**	9 × 4
6	3 × 9	**13**	10 × 5
7	6 × 6	**14**	8 × 6

EXERCISE 1.10B

Copy these into your book and write down the answers.

1	2 × 3	**9**	2 × 2
2	7 × 6	**10**	3 × 7
3	4 × 8	**11**	6 × 4
4	9 × 9	**12**	8 × 8
5	5 × 4	**13**	10 × 3
6	4 × 3	**14**	5 × 7
7	9 × 6	**15**	8 × 9
8	8 × 5		

Unlike multiplication, there are different ways of writing divisions.

$42 \div 7$, $\frac{42}{7}$ and $7\overline{)42}$

are three ways to say 'divide 42 by 7.'

A multiplication grid can be helpful but knowing your tables by heart is much better.

x	1	2	3	4	5	6	7	8	9	10
1	1	2	3	4	5	6	7	8	9	10
2	2	4	6	8	10	12	14	16	18	20
3	3	6	9	12	15	18	21	24	27	30
4	4	8	12	16	20	24	28	32	36	40
5	5	10	15	20	25	30	35	40	45	50
6	6	12	18	24	30	36	42	48	54	60
7	7	14	21	28	35	42	49	56	63	70
8	8	16	24	32	40	48	56	64	72	80
9	9	18	27	36	45	54	63	72	81	90
10	10	20	30	40	50	60	70	80	90	100

For 42 ÷ 7 you look for 42 on the 7 row. Directly above the 42 shows the answer. So 42 ÷ 7 = 6.

ACTIVITY 4

Copy and complete this multiplication table.

×	2	5	3	8	6
4					
2					
5					
9					
6					

ACTIVITY – EXTENSION

Make up a multiplication table of your own like the one above. See if a friend can complete it.

ACTIVITY 5

1 Multiply two even numbers together.
Do this several times with different even numbers.
What do you notice about your answers?

2 Multiply two odd numbers together.
Do this several times with different numbers.
What do you notice about your answers?

3 Multiply one odd and one even number together.
Do this several times with different odd and even numbers.
What do you notice about your answers?

EXERCISE 1.11A

Work out

1 $48 \div 6$

2 $4\overline{)28}$

3 $\frac{12}{3}$

4 $36 \div 9$

5 $54 \div 6$

6 $\frac{16}{2}$

7 $7\overline{)63}$

8 $35 \div 7$

9 $\frac{72}{8}$

10 $48 \div 8$

EXERCISE 1.11B

Work out

1 $14 \div 2$

2 $9\overline{)63}$

3 $\frac{21}{7}$

4 $70 \div 10$

5 $\frac{24}{4}$

6 $27 \div 9$

7 $40 \div 5$

8 $\frac{18}{2}$

9 $8\overline{)64}$

10 $\frac{45}{5}$

Just as with addition and subtraction, multiplication and division are linked together.

EXAMPLE 7

$42 \div 7 = 6$ $42 \div 6 = 7$ $6 \times 7 = 42$

Knowing one answer helps you work out the answers to the others.

EXERCISE 1.12A

Use the following numbers to show three different × or ÷ calculations.

1 5   7   35   **4** 6   48   8
2 63   9   7   **5** 9   36   4
3 24   6   4

EXERCISE 1.12B

Use the following numbers to show three different × or ÷ calculations.

1 7   4   28   **4** 8   72   9
2 9   45   5   **5** 6   18   3
3 8   2   4

Key ideas

● When rounding numbers up or down, a 5 always rounds up.
● To identify odd and even numbers look at the final digit of the number.
● To identify multiples of 5 and 10, look at the final digit of the number.
● You should know the times tables up to 10×10.
● Addition and subtraction sums are linked.
● Multiplication and division sums are linked.
● There are strategies for tackling addition and multiplication sums.

2 Probability

Chance

What is the chance of it raining sometime in the next three weeks?

If that question is asked in the UK, then the answer is probably 'very likely', even if it is the middle of summer! In the UK climate, even in summer, there is often some rain over a period as long as three weeks.

In many other countries the answer will be very different. In extremely dry areas of the world the answer may be 'very unlikely'. In others the answer may be 'about evens'. By 'evens' we mean there is an equal chance of something happening as not happening. In some countries the answer may well depend on what season it is when the question is asked, as long periods of wet weather are often followed by long periods of dry weather.

So, wherever the question is asked, the answer will be on a scale like the one below.

| impossible | very unlikely | unlikely | evens | likely | very likely | certain |

EXAMPLE 1

Use arrows to place the chances of the following events on a scale.

a) the chance of a coin coming down 'heads' when it is tossed

b) the chance of a cow jumping over the moon

c) the chance of getting mathematics homework this week

This works out as follows.

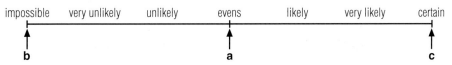

Note: c) could be variable but should be at least this far to the right!

The word 'probability' is used instead of the word 'chance' in mathematics.

EXAMPLE 2

Use arrows to place the probability of the following events on a scale.

a) the probability of getting an odd number when a fair die is thrown

b) the probability that it will get dark tonight

c) the probability that United will score five goals in their next match

Answer

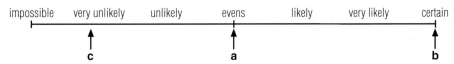

Note: that the position of **c)** is still placed largely through guesswork.

EXERCISE 2.1A

1   Make a copy of this scale.

impossible   very unlikely   unlikely   evens   likely   very likely   certain

Place and label arrows on your scale to show:
a) the chance that Christmas Day is on 25 December this year
b) the chance that it will rain every day in June
c) the chance that the next baby born at your local hospital is a girl
d) the chance that the next lorry you see will have a male driver
e) the chance that the traffic lights are showing blue.

2   Make a copy of this scale.

impossible   evens   certain

Place and label arrows on your scale to show:
a) the probability of getting tails when you throw a coin
b) the probability that a football team from the top division will win the cup
c) the probability that you will get a seven when you throw an ordinary fair die
d) the probability that it will snow in England on 1 June
e) the probability that a person selected at random from your school is
      left handed.

3   Use the probability words 'Unlikely ... to ... Certain' to answer the following
      questions about rolling a normal six-sided die.
a) I will get a 0.
b) I will get a 3, 2 or 1.
c) I will get a 6 or less.
d) I will get an even number.
e) I will get a 4 or 5.

Exercise 2.1A cont'd

4 **a)** A box contains 12 counters. The counters are green, blue or yellow. A counter is chosen from the box at random. If there is an evens chance of getting a green counter, how many green counters are there?

   b) Another box contains 12 counters. There are 5 red counters and 7 black counters. One counter is chosen from the box without looking.
   (i) Which colour is it more likely to be? Explain why.
   (ii) How many of each colour should be removed to make it an evens chance that either colour is chosen?

5 **a)** Here is a spinner.
   (i) Draw and colour the spinner so that there is an even chance of it landing on black or white.

   (ii) How many different spinners can you draw so that the chance of landing on black or white is evens?

   b) Here is another spinner.
   For each part of the question, draw and colour a spinner to answer the question.
   (i) It is certain to land on black.
   (ii) It is unlikely to land on black.
   (iii) It is impossible to land on black.
   (iv) It is very likely to land on black.

EXERCISE 2.1B

1 Make a copy of this scale.

| impossible | very unlikely | unlikely | evens | likely | very likely | certain |

Place and label arrows on your scale to show
a) the chance that the next vehicle to go past the school will be a car
b) the chance that a ball thrown in the air will come down
c) the chance that you will get a car for your seventeenth birthday
d) the chance that a cow will jump over the moon
e) the chance that the sun will be shining when you wake up tomorrow.

2 Make a copy of this scale.

| impossible | evens | certain |

Place and label arrows on your scale to show:
a) the probability that you will get maths homework tonight
b) the probability that the winning numbers in the next National Lottery draw are 1, 15, 23, 29, 35, 46
c) the probability that this lesson will finish before 6 p.m.
d) the probability that the next car that passes the school will have four passengers
e) the probability that when you cut a pack of ordinary playing cards you get a red card.

3 Use the probability words 'Unlikely ... to ... Certain' to answer the following questions about choosing a card from an ordinary pack of playing cards.
a) I will get a red card.
b) I will get a blue card.
c) I will get an ace.
d) I will get a Heart, Spade, Diamond or Club.
e) I will get a number card.

Exercise 2.1B cont'd

4 a) I have the eight cards shown,
two are already turned over.
I turn them all over, shuffle them
and then pick one. What numbers
would the unknown cards be if

   (i) the chance of picking a 1 is evens

   (ii) the chance of picking a 1 or a 2 is the same.

b) Write down two events which you think have a very likely chance of
happening tomorrow, and two which are unlikely to happen.

5 Below are pictures of four jars.
Each jar contains black and white beads only.
Make a drawing of each of the jars and its beads in your book.
For each of the jars, colour the beads necessary for the probability to be correct.

a)

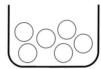

The chance of choosing
a black is certain.

b)

There is an evens chance
of choosing a black.

c)

The chances of choosing
a black is unlikely.

d)

The chance of choosing
a black is very likely.

'The probability washing line'

Put a piece of string, 3 metres long, across the front of the classroom, either from wall to wall or across the face of the front wall, Write the words 'Impossible, Very Unlikely, Unlikely, Evens, Likely, Very likely, Certain, on pieces of paper and attach them at equal intervals along the string. Each member of the class writes on a piece of paper some words about an event that has not yet happened, such as 'Tomorrow it will rain'. Decide as a class where the pieces of paper should go.

Key ideas

- You should know how to use the vocabulary of probability.
- You should know the words 'event' and 'chance'.
- You should be able to use the probability scale.

A1 Revision exercise

1   Make as many different numbers as you can with the digits 4, 9, 7.

2   **a)** Write in figures twenty-five thousand and thirty-seven.

   b) Write in words 1 604 719.

3   Round 7835:

   a) to the nearest 10

   b) to the nearest 100

   c) to the nearest 1000.

4   List all the odd numbers between 16 and 28.

5   List all the even numbers between 121 and 137.

6   List all the numbers between 61 and 93 that are divisible by 5.

7   List all the numbers between 207 and 315 that are divisible by 10.

8   Work out

   a) 56 + 8

   b) 23 + 44

   c) 37 + 15

   d) 64 + 36

   e) 52 + 28

   f) 67 − 5

   g) 95 − 14

   h) 62 − 23

   i) 84 − 27

   j) 73 − 65

9   Work out

   a) 2×3

   b) 7×5

   c) 6×6

   d) 4×9

   e) 8×7

   f) $30 \div 6$

   g) $9 \overline{)\,27}$

   h) $\frac{80}{10}$

   i) $45 \div 5$

   j) $\frac{32}{4}$

10 Copy this scale.

Put an arrow on the scale to show the chances:

   a) of you going to school on Christmas Day

   b) of a horse winning the next Grand National

   c) that somebody in your class will get more than 50% in the next mathematics test.

11 Copy this scale.

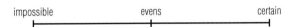

Place and label arrows on your scale to show the probability that:

   a) a coin will come down heads when it is tossed

   b) it will rain sometime in the next two weeks

   c) the school team will win their next hockey match 7–0.

12 You have an eight-sided spinner numbered 1 to 8. Use the probability words 'Impossible, Very unlikely ... Very likely, Certain' to describe the chance of the following things happening when it lands.

a) an odd number,

b) a number less than 7

c) the number 8

d) the number 10

13 The six-sided spinner shown in the diagram has spotty, striped or plain sections.

a) Write down, from most likely to least likely, the order of the sections on which the spinner will land. Give a reason for your answer.

b) Which section of the spinner should be changed to make the chance of it landing on any of the types evens?

3 *Direction and position*

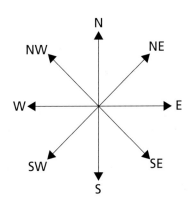

The diagram shows the main compass directions.

EXAMPLE 1

This is a sketch map of an Island. North in marked.

a) Which letter is North of C?

b) Which letter is West of C?

c) What is the direction from A to B?

d) What is the direction from F to D?

a) A **b)** F **c)** South West (SW) **d)** North East

N

•A   E •

B •

•   •C   D •
F

EXERCISE 3.1A

Where a North line is required, draw it pointing up the page.

1 Mark two points, A and B, where AB is 4 cm long and A is West of B.

2 a) Mark three points, A, B and C, where A is 4 cm North of B and C is 4 cm west of B.

b) What is the direction AC?

3

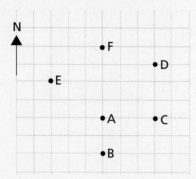

Name the letter

a) South of A

b) North East of A

c) South West of C.

4

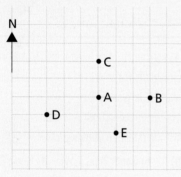

Write down the directions of

a) A to B

b) C to D

c) E to B.

5

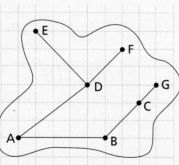

a) Write down the directions of
   (i) A to B   **(ii)** B to C.

b) What is
   (i) South East of F
   (ii) North East of C?

EXERCISE 3.1B

Where North lines are required, draw them pointing up the page.

1 Mark two points, B and C, where BC is 3 cm long and C is South West of B.

2 **a)** Mark three points, A, B and C, where A is 2 cm West of B and C is 2 cm South of A.

b) What is the direction CB?

3

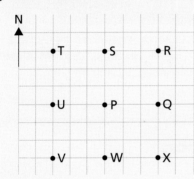

Name the letter

a) South of P

b) North West of W

c) South East of P.

4

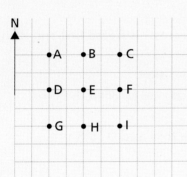

Write down the directions of

a) A to B

b) C to G

c) E to B

d) I to E.

5 Monyash is South East of Flagg. What is the direction from Flagg from Monyash?

ACTIVITY 1

Look at the last diagram above.

- How many ways are there to get from A to E?
  List them, e.g. 'Go East to B, then go South to E'.
- How many ways are there to get from G to F?

ACTIVITY 2

Mark the four main compass points on the walls of the classroom.

Take it in turns to direct a student around the room using compass directions only.

ACTIVITY 3

Get an Ordnance Survey map of the local area. Find and write down the compass directions between different places on the map.

Graphs, coordinates and direction

Coordinates

To find a point on a grid, the **horizontal** position (across) is stated first and then the **vertical** (upward) position.

On this grid two lines are drawn. These are called the **axes**.

The horizontal one is called the *x*-axis and the vertical one is called the *y*-axis.

The lines of the grid are numbered from 0 to 8 across and 0 to 8 up.

Rather than putting two 0 where the two axes cross, one 0 is placed there.

This point is called the **origin**.

The point marked A is on the line 2 across and 3 up. This is written as (2,3). These are the **coordinates of A**.

The point marked B is 5 across and 6 up, so its coordinates are (5,6).

The origin is 0 across and 0 up so its coordinates are (0,0).

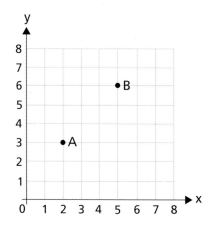

EXAMPLE 2

State the coordinates of the points marked P, Q and R on the grid.

P is (1,3)   Q is (5,1)   R is (3,0)

Notes: Take care with the last one, it is 3 across and 0 up, so it is on the *x*-axis.

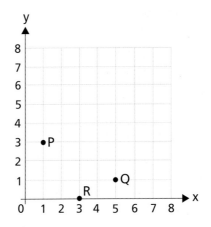

EXAMPLE 3

On the grid, mark and label the points A (8,4)
B (4,7)   C (0,6).

The points can be plotted with a dot or a cross, and the letter written next to them.

The hard point to plot is the last one which is 0 across and 6 up, so it is on the *y*-axis.

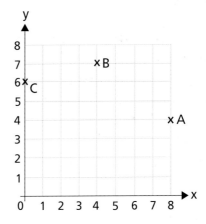

EXAMPLE 4

a) State the coordinates of the points marked D, E and F on the grid.

b) Mark the points G (4,8)   H (6,11)   I (7,15).

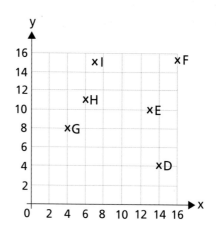

a) D is (14,4)   E is (13,10)   F is (16,15).

The lines are marked 2, 4, 6, etc., so 13 is halfway between 12 and 14.

b) The points are plotted on the grid.

Exam tip

A common error when plotting points is to label the spaces instead of the lines.

Exam tip

A common error when reading off graphs is to confuse, for example, (0,2) and (2,0).

EXERCISE 3.2A

1 Write down the coordinates of the points marked A, B and C on the grid.

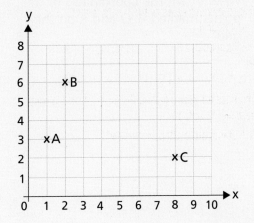

2 Write down the coordinates of the points marked D, E and F on the grid.

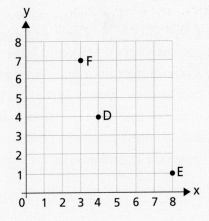

3 Write down the coordinates of the points marked L, M and N on the grid.

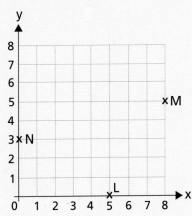

Chapter 3   *Direction and position*

Exercise 3.2A cont'd

4 Write down the coordinates of the points marked P, Q and R on the grid.

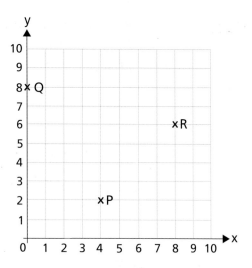

5 Draw a pair of axes on a grid and mark the lines 0 to 8 for both *x* and *y* (as in the diagrams above). Mark and label the points A (4,7)   B (1,3)   C (5,8).

6 Draw a pair of axes on a grid and mark the lines 0 to 8 for both *x* and *y*.
Mark and label the points S (5,3)   T (7,2)   W (0,5).

7 Draw a pair of axes on a grid and mark the lines 0 to 8 for both *x* and *y*.
Mark and label the points M (2,8)   N (7,7)   R (5,0).

8 Write down the coordinates of the points marked A, B, C and D on the grid below.

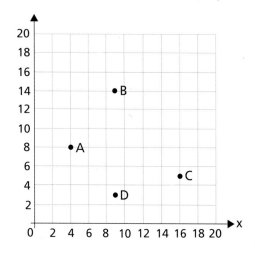

Exercise 3.2A cont'd

9 Draw a pair of axes on a grid and mark the lines 0 to 8 for both *x* and *y*.

   a) Mark and label the points A (1,4) and B (5,4).

   b) Join the points A and B with a straight line.

   c) Write down the coordinates of the midpoint of the line AB.

10 Draw a pair of axes on a grid and mark the lines 0 to 20, in twos, for both *x* and *y*.

   a) Mark and label the points
   C (2,18)   D (2,7)   E (9,18).

   b) Join the three points to make a triangle.

   c) What is special about the triangle?

EXERCISE 3.2B

1 Write down the coordinates of the points marked A, B and C on the grid.

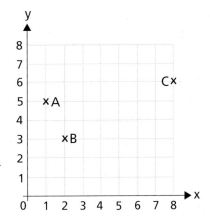

Exercise 3.2B cont'd

2 Write down the coordinates of the points marked D, E and F on the grid.

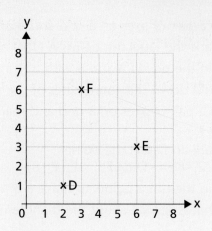

3 Write down the coordinates of the points marked G, H and I on the grid.

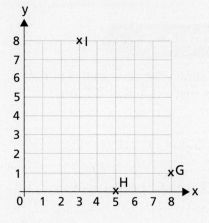

4 Write down the coordinates of the points marked P, Q and R on the grid.

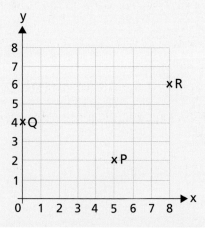

Exercise 3.2B cont'd

5 Draw a pair of axes on a grid and mark the lines 0 to 6 for both *x* and *y* (as in the diagrams above). Mark and label the points A (5,1)   B (2,3)   C (4,6).

6 Draw a pair of axes on a grid and mark the lines 0 to 8 for both *x* and *y*. Mark and label the points S (1,3)   T (8,2)   W (4,0).

7 Draw a pair of axes on a grid and mark the lines 0 to 10 for both *x* and *y*. Mark and label the points M (2,5)   N (4,7)   P (0,8)   R (9,10).

8 Write down the coordinates of the points marked A, B, C and D on the grid.

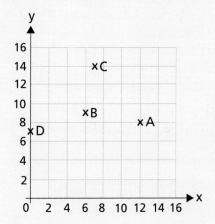

9 Draw a pair of axes on a grid and mark the lines 0 to 8 for both *x* and *y*.

a) Mark and label the points A (3,4) and B (3,8).

b) Join the points A and B with a straight line.

c) Write down the coordinates of the midpoint of the line AB.

10 Draw a pair of axes on a grid and mark the lines 0 to 20 in twos, for both *x* and *y*.

a) Mark and label the points A (5,12)   B (15,6)   C (4,2) and D (16,16).

b) Join AB and CD.

c) Write down the coordinates of the point where the two lines cross.

ACTIVITY 4

'Cover Up'

This is a game for two students. You need two dice of different colours, e.g. one red dice and one blue dice. Draw out a grid like the one here.

Roll the two dice.

The score on the blue die will give the x-coordinate and the score on the red die the y-coordinate of a point. Put a cross at the point if you are Player 1 or a circle if you are Player 2.

Continue, in turn, to throw the dice and mark points until all of the grid has been covered.

The player with the most crosses or circles wins.

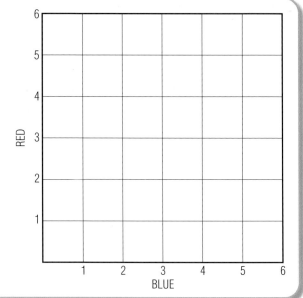

ACTIVITY – EXTENSION

Play the game as before. However, this time, if you get a point that is already covered, then you lose your turn!

Key ideas

- There are 8 main compass directions.
- You should know the name of the x-axis, the y-axis and the origin.
- You should be able to locate a point on a grid, given its coordinates.
- You should know how to write down the coordinate of a point (across, up) or (x,y).

4 Scales

Telling the time

Reminder: The two clock faces below show what the minute hand (the long hand) tells you when it points to the numbers on the clock.

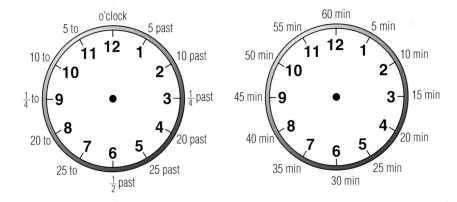

EXAMPLE 1

Write the time shown on the clocks in words and figures.

a)

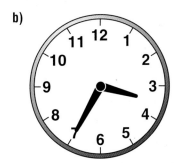

five past six

6:05

b)

twenty five to four

3:35

EXERCISE 4.1A

Write these times in words and figures.

1

2

3

4

5

6
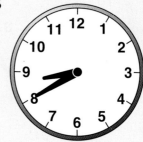

7

8
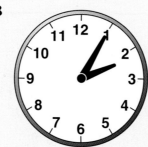

9

10

For each of the following questions, use the worksheet or copy a clock face into your book and mark on it the time given.

11 10 past 6   **12** $\frac{1}{4}$ past 2   **13** 5 o'clock

14 6:50   **15** 10:35   **16** 25 past 7

17 10 to 12   **18** $\frac{1}{4}$ to 3   **19** 3:45

20 12:40   **21** 25 to 6   **22** 1:55

EXERCISE 4.1B

Write these times in words and figures.

1

2

3

4

5

6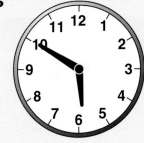

7

8

9

10

For each of the following questions, use the worksheet or copy a clock face into your book and mark on it the time given.

11 25 to 5 **12** 12 o'clock **13** $\frac{1}{2}$ past nine

14 11:20 **15** 7:45 **16** 20 past 2

17 $\frac{1}{4}$ to 4 **18** 6:55 **19** $\frac{1}{2}$ past 5

20 11:05 **21** 10 to 8 **22** 9:50

How long does it take?

EXAMPLE 2

How many minutes are there between the following times?

a) 7:25 to 7:50

b) $\frac{1}{4}$ to 10 to 25 past 10

a) From 25 minutes to 50 minutes is 25 minutes.

b) From $\frac{1}{4}$ to 10 up to 10 o'clock   15 mins

     from 10 o'clock to 25 past 10   <u>25 mins +</u>

                                               <u>40 mins</u>

EXERCISE 4.2A

How many minutes are there between the following times?

Use the clock faces at the beginning of the chapter to help you.

	Start time	Finish time
1	8:00	8:15
2	6:20	6:50
3	4:10	4:35
4	$\frac{1}{4}$ past 3	20 to 4
5	$\frac{1}{4}$ to 10	5 to 10
6	2:30	3:10
7	11:55	12:20
8	$\frac{1}{4}$ to 7	20 past 7
9	1:40	2:35
10	10 to 4	25 to 5

EXERCISE 4.2B

How many minutes are there between the following times?

Use the clock faces at the beginning of the chapter to help you.

	Start time	Finish time
1	2:30	2:55
2	10:10	10:45
3	3:05	4:00
4	$\frac{1}{4}$ past 11	$\frac{1}{4}$ to 12
5	20 past 7	20 to 8
6	2:50	3:15
7	8:20	9:05
8	4:35	5:15
9	$\frac{1}{2}$ past 10	5 past 11
10	25 to 9	$\frac{1}{2}$ past 10

Finishing times

EXAMPLE 3

Work out the finishing time for each of the following:

a) starts at 1:35 and lasts for 15 minutes.

b) starts at $\frac{1}{2}$ past 9 and lasts for 45 minutes.

a) Add 15 on to 35 to give a finishing time of 1:50.

b) Add 30 of the 45 minutes to take it up to 10 o'clock.
Add on the remaining 15 minutes, giving a finishing time of 10:15.

EXERCISE 4.3A

1 Copy and complete the table to find the finishing times.

	Starts	Lasts	Finishes
a)	2:05	30 mins	
b)	4:15	50 mins	
c)	11:35	25 mins	
d)	5 to 9	45 mins	

2 A plane leaves Heathrow at 10:05 and flies to Leeds. The flight lasts 45 minutes. At what time does the plane land?

3 Break time starts at 10:25 and lasts for 20 minutes. At what time does school re-start?

4 The TV news starts at $\frac{1}{4}$ to 9 and lasts for 20 minutes. At what time does the news finish?

5 A boy starts delivering newspapers at 6:55. He finishes 55 minutes later. At what time does he finish?

EXERCISE 4.3B

1 Copy and complete the table to find the finishing times.

	Starts	Lasts	Finishes
a)	6:10	35 mins	
b)	8:55	20 mins	
c)	5:30	15 mins	
d)	20 to 1	15 mins	
e)	$\frac{1}{4}$ past 10	25 mins	

2 My favourite TV programme lasts for 45 minutes. I start watching it at 7:05. At what time does it finish?

3 A train leaves Chesterfield Station at 20 past 1. It arrives in Sheffield 20 minutes later. What time does it arrive?

4 A rugby match starts at 2:40. The first half lasts for 35 minutes. At what time does the first half finish?

5 My bus to work takes 15 minutes. What time will I arrive at work if I catch the bus at 8:50?

Reading scales

To read a scale, decide what each small division is and read it from there.

EXAMPLE 4

What are the readings at A and B on the scale below?

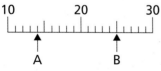

A 14   B 25   Here there are 10 divisions between 10 and 20 so each small division is 1. If the marked lines had been 1, 2 and 3 instead of 10, 20 and 30, each small division would have been 0·1 so A would be 1·4 and B would be 2·5.

EXAMPLE 5

a) What is the reading at C on the scale below?

b) Mark 63 with a letter D.

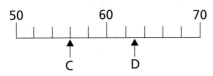

a) C 56   Here there are 5 divisions between 50 and 60, so each is 2.

b) D marked at 63   The divisions are at 62 and 64, so D is midway between.

EXAMPLE 6

a) Read the point on the scale marked E.

b) Mark 650 with the letter F.

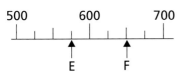

a) E 575   Here there are four divisions between 500
                      and 600, so each is 25.

b) F marked at 650.

Exam tip

Most errors in reading scales are made because the smallest division is assumed to be 0·1, 1 or 10, etc. Make sure you know what the smallest division is before you start reading or marking on a scale.

EXERCISE 4.4A

Where you are asked to mark a reading make a sketch of the scale and mark the point on it.

1 Read the points marked A and B on the scale below.

2 Read the points marked C and D on the scale below.

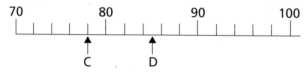

3 Read the points marked E and F on the scale below.

4 **a)** Read the point marked G on the scale below.
   b) Mark H at 37.

5 **a)** Read the point marked I on the scale below.
   b) Mark J at 250.

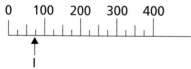

6 **a)** Read the point marked K on the scale below.
   b) Mark L at 160.

Exercise 4.4A cont'd

7 a) Read the point marked M on the scale below.

b) Mark N at 700.

8 What is the reading on this thermometer?

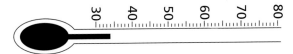

9 What is the reading on this dial?

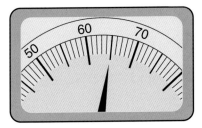

10 How much does this weigh?

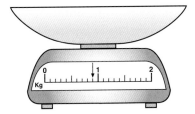

EXERCISE 4.4B

1   Read the points marked A and B on the scale below.

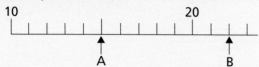

2   Read the points marked C and D on the scale below.

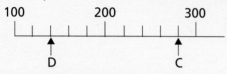

3   Read the points marked E and F on the scale below.

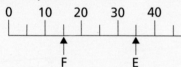

4   Draw a sketch of the scale below and mark G at 48 and H at 62.

5   **a)** Read the point marked I on the scale below.

   b) Draw a sketch of the scale and mark J at 0·3.

6   **a)** Read the point marked K on the scale below.

   b) Draw a sketch of the scale and mark L at 63.

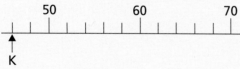

Exercise 4.4B cont'd

7 **a)** Read the points marked M and N on the scale below.

b) Draw a sketch of the scale and mark Q at 105.

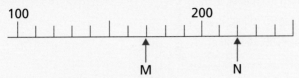

8 What is the reading on this dial?

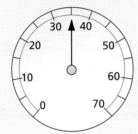

9 What is the reading on this speedometer?

10 a) What is this reading?

b) Alan finds the accurate pressure from the internet. It is 29·3. Mark this on a copy of the diagram.

ACTIVITY

Design the dashboard of a car.

Include:
- a speedometer going from 0 to 100 mph in 5 s
- a fuel gauge going from 0 to 20 gallons in 2 s
- a temperature scale going from ⁻10 to 30 in 1 s.

Key ideas

- You should know how to tell the time and write time in words and figures.
- There are strategies to find time intervals and finishing times.
- There are strategies for reading scales in different graduations.

Revision exercise

1 •A   •B   •C

   •D   •E   •F

   •G   •H   •I

a) What is the direction of

    (i)   A to B   **(ii)** B to D

    (iii) H to E   **(iv)** G to E

    (v)   I to H

b) Name the letter that is

    (i) North of E **(ii)** South East of B

2 John walked along a road that went South West. David walked the other way along the same road. In what direction did David walk?

3 Write the coordinates of the points marked on the grid below.

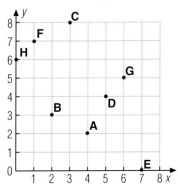

4 Draw axes like the ones in Question 3. Plot and label the points

I (1,4)   J (8,2)   K (0,6)
L (5,7)   M (2,1)   N (3,5)
P (4,0) and Q (7,3)

5 Write the coordinates of the points marked on the grid below.

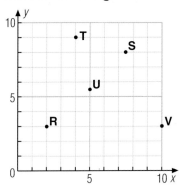

6 Copy out this grid. Plot and label the points W (1,3)   X (3,7)   Y $(2\frac{1}{2},5)$ Z $(4\frac{1}{2},8\frac{1}{2})$.

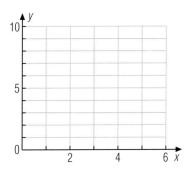

7 Write the time shown on these clocks in words and figures.

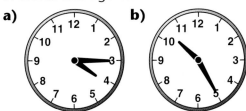

c)

8 The late evening news starts at $\frac{1}{4}$ past 11 and lasts for 20 minutes. At what time does it finish?

9 School finishes at 3:15 p.m. I arrive home at 4:05 p.m. How long has it taken me to get home?

10 The train takes 35 minutes to travel from Nottingham to Leicester. It leaves Nottingham at 2:45. What time does it arrive in Leicester?

11 Write down the points marked on these scales.

a)

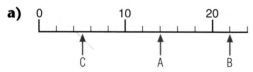

b)

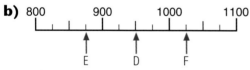

c)

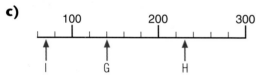

12 a) Copy this scale and mark on the points A 3·5   B 1·8   C 2·4.

b) Copy this scale and mark on the points D 140   E 185   F 205 G 162.

c) Copy this scale and mark on the points H 45   I 72   J 89.

```
40      50      60      70      80      90
|___|___|___|___|___|___|___|___|___|___|
```

5 Algebra patterns and using letters for numbers

Letters for numbers

This line is made up of two pieces, 6cm and 4cm long.

You can see that the full length is 6 + 4 = 10cm.

If the second piece is 8cm the full length would be 6 + 8 = 14cm.

If the second piece is some length that is not known, a letter can be used for the length.

Here the length has been called x.

The full length is $6 + x$ or $x + 6$. This is called an **expression** for the length.

When a letter is used instead of a number it is called **algebra**.

Here the letter is x but any letter can be used.

In these examples and exercises the lines are not drawn accurately so do not measure them.

All lengths are in centimetres.

EXAMPLE 1

a) Write down the length of the full line.

b) Write down the expression for the length, in cm, of each of these lines.

(i)

(ii)

a) Length is 2 + 5 = 7 cm.

b) (i) Length is 2 + q.
  (ii) Length is 3 + y.

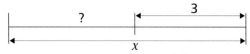

This line is x cm long altogether. One part is 3 cm, so an expression for the length, in cm, of the other part is x − 3.

EXAMPLE 2

Write down expressions for the length of the part of the lines marked with a ?.

a)

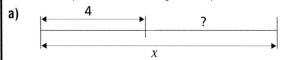

b)

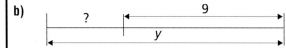

c)

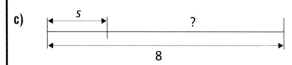

a) ? is x − 4.   **b)** ? is y − 9.   **c)** ? is 8 − s.

If two lines, each x cm long, are put end to end, the expression for the full length is $x + x$ or $2 \times x$, and this can be written as $2x$.

EXAMPLE 3

a) Write down expressions for the lengths of these lines.

(i)
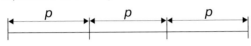

(ii)
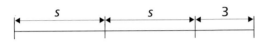

b) Write down expressions for the lengths of the parts of the lines marked ?.

(i)
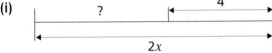

(ii)
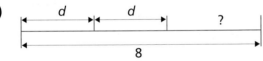

a) (i) Length is $p + p + p = 3p$
   (ii) Length is $s + s + 3 = 2s + 3$

b) (i) Length of ? is $2x - 4$
   (ii) Length of ? is $8 - 2d$

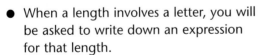

Exam tip

- When a length involves a letter, you will be asked to write down an expression for that length.
- $p + 5$, $t - 3$, $8 - x$ and other similar expressions cannot be written more simply, but $p + 3$ is the same as $3 + p$.
- $n + n$ or $2 \times n$ can be written as $2n$.

Chapter 5 *Algebra patterns and using letters for numbers*

EXERCISE 5.1A

1 a) Write down the length of the line below.

b) Write down expressions for the lengths of the lines.

(i)

(ii)

2 a) Write down the length of the line.

b) Write down expressions for the lengths of the lines.

(i)

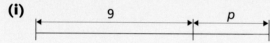

(ii)

3 Write down expressions for the lengths of the lines.

a)

b)

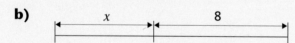

4 a) Write down the length of the line.

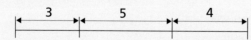

b) Write down an expression for the length of the line.

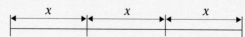

Exercise 5.1A cont'd

5 Write down expressions for the lengths of the lines.

a)
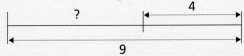

b)
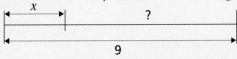

6 a) Write down the length marked *?* on the line.
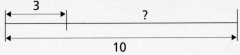

b) Write down an expression for the length marked *?* on the line.
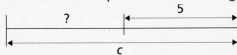

7 a) Write down the length marked *?* on the line.
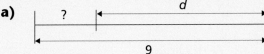

b) Write down an expression for the length marked *?* on the line.
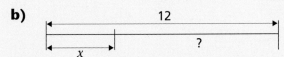

8 Write down expressions for the lengths marked *?* on the lines.

a)
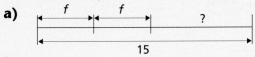

b)
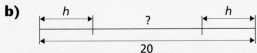

9 Write down expressions for the lengths marked *?* on the lines.

a)

b)

Chapter 5 *Algebra patterns and using letters for numbers*

Exercise 5.1A cont'd

10 Write down expressions for the lengths marked *?* on the lines.

a)

b)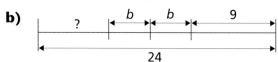

EXERCISE 5.1A

1 a) Write down the length of the line.

b) Write down expressions for the lengths of the lines.

(i)

(ii)

2 a) Write down the length of the line.

b) Write down expressions for the lengths of the lines.

(i)

(ii)

Exercise 5.1B cont'd

3   Write down expressions for the lengths of the lines.

a)

b)

4   **a)** Write down the length of the line.

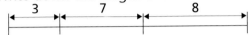

b) Write down an expression for the length of the line.

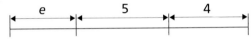

5   Write down expressions for the lengths of the lines.

a)

b)

6   Write down expressions for the lengths of the lines.

a)

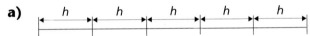

b)

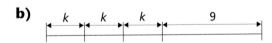

7   Write down expressions for the lengths marked *?* on the lines.

a)

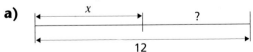

b)

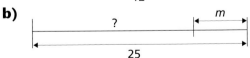

Exercise 5.1B cont'd

8 Write down expressions for the lengths marked *?* on the lines.

a)

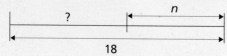

b)

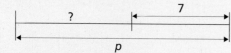

9 Write down expressions for the lengths marked *?* on the lines.

a)

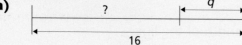

b)

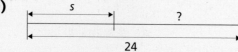

10 Write down expressions for the lengths marked *?* on the lines.

a)

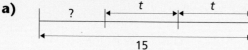

b)

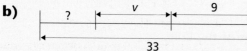

John is 5 years older than David.

When David was 4, John was 4 + 5 = 9.

When David is 10, John will be 10 + 5 = 15.

You can see that John's age will always equal David's age plus 5.

Algebra helps you to write this by using a letter for David's age.

Any letter can be used. So let David's age be *d*.

Then the expression for David's age is *d* + 5.

EXAMPLE 4

Margaret has 3 more CDs than Andrea.

a) Write down the number of CDs that Margaret has when Andrea has
  (i) 3   **(ii)** 12.
b) Write an expression for the number of CDs that Margaret has when Andrea has m.

a) (i) $3 + 3 = 6$ **(ii)** $12 + 3 = 15$
b) $m + 3$.

EXAMPLE 5

There are 6 fewer boys than girls in form 10A.

a) Write down how many boys there are when there are
  (i) 12 girls   **(ii)** 17 girls.
b) Write down an expression for how many boys there are when there are g girls.

a) (i) $12 - 6 = 6$ **(ii)** $17 - 6 = 11$
b) $g - 6$.

If Sarah was given 20 pence on Monday and another 20 pence on Tuesday she would have $20 + 20$ or 2×20 = 40 pence altogether.

If she was given q pence on Monday and another q pence on Tuesday she would have $q + q$ or $2 \times q$ pence. A short way to write $2 \times q$ is $2q$.

EXAMPLE 6

A kilo of apples costs a pence. Write down an expression for the cost of

a) 2 kilos   **b)** 6 kilos
c) 10 kilos.

a) 2 kilos cost $2 \times a = 2a$ pence.
b) 6 kilos cost $6 \times a = 6a$ pence.
c) 10 kilos cost $10 \times a = 10a$ pence.

EXERCISE 5.2A

1 The swimming club has 5 more boys than girls.
   a) Write down the number of boys if there are   **(i)** 10 girls   **(ii)** 18 girls.
   b) Write down an expression for the number of boys if there are g girls.

2 Fiona swam 10 more lengths than Dan.

   a) Write down the number of lengths that Fiona swam if Dan swam
      (i) 15   **(ii)** 32 lengths.
   b) Write down an expression for the number of lengths that Fiona swam if Dan swam L lengths.

3 Holly spent £5 more than Ross at the shops.
   a) Write down the amount that Holly spent if Ross spent   **(i)** £24   **(ii)** £38.
   b) Write down an expression for the amount that Holly spent if Ross spent £C.

4 Ann has 12 more photos than Pat.
   a) Write down the number of photos Ann has if Pat has
      (i) 6 photos **(ii)** 24 photos.
   b) Write down an expression for the number of photos Ann has if Pat has p photos.

5 The length of a rectangle is 6 cm longer than its width.
   a) Write down the length of the rectangle if the width is   **(i)** 12 cm   **(ii)** 4 cm.
   b) Write down an expression for the length of the rectangle if the width is w cm.

6 Mr Lewis earns £80 less per week than Mr Thompson.
   a) How much does Mr Lewis earn if Mr Thompson earns **(i)** £180   **(ii)** £250?
   b) Write down an expression for the amount that Mr Lewis earns if Mr Thompson earns £x.

Exercise 5.2A cont'd

7 Carol scored 14% less in her exam than Sui.

a) What percentage did Carol score if Sui scored   **(i)** 85%   **(ii)** 43%?

b) Write down an expression for the percentage Carol scored if Sui scored m%.

8 Asma has t pound coins in her purse.

Write down expressions for the number of pound coins that Beccy has if she has

a) 5 less   **b)** 9 less   **c)** 8 more than Asma.

9 Cheese costs £4 a kilogram.

a) How much does it cost for   **(i)** 2 kilograms   **(ii)** 5 kilograms?

b) Write down an expression for the cost of c kilograms.

10 Sian spends twice as much time on homework as Leslie.

a) How many minutes does Sian spend on homework if Leslie spends
    (i) 20 minutes   **(ii)** 45 minutes?

b) Write down an expression for the number of minutes Sian spends on
    homework if Leslie spends m minutes.

11 The length of a rectangle is twice as long as its width.

a) How long is the rectangle if the width is   **(i)** 4 cm   **(ii)** 3·2 cm?

b) Write down an expression for the length of the rectangle if the width is w cm.

12 Joe buys a chocolate bar costing x pence every day.

Write down expressions for the amount he spends in

a) 3 days   **b)** 5 days   **c)** 7 days.

EXERCISE 5.2B

1 Pam has 4 more books than James.

   a) Write down the number of books that Pam has if James has
   (i) 12   **(ii)** 20 books.

   b) Write down an expression for the number of books that Pam has if James
   has x books.

2 Coffee costs 15 pence more per cup than tea.

   a) Write down the cost of a cup of coffee if a cup of tea costs
   (i) 20 pence   **(ii)** 45 pence.

   b) Write down an expression for the cost of a cup of coffee if a cup of tea costs
   t pence.

3 Linford has 8 more toy cars than Thomas.

   a) How many toy cars has Linford if Thomas has   **(i)** 12 cars   **(ii)** 18 cars?

   b) Write down an expression for the number of cars that Linford has if Thomas
   as c cars.

4 There are b more bananas on a fruit stall than oranges.

   Write down an expression for the number of bananas if the number of oranges is

   a) 5   **b)** 8   **c)** 9.

5 Stephen is 4 cm shorter than Alan.

   a) How tall is Stephen if Alan's height is   **(i)** 136 cm   **(ii)** 152 cm?

   b) Write down an expression for Stephen's height if Alan's height is q cm.

6 There were 42 passengers on the
   bus when it arrived at the stop.
   Some got off.

   a) How many were left on the bus
   if the number who got off was
   (i) 4   **(ii)** 23?

   b) Write down an expression for
   the number left on the bus
   if the number who got off
   was p.

Exercise 5.2B cont'd

7 Toni weighs 2 kg less than Peter.

    a) How much does Toni weigh if Peter weighs   **(i)** 56 kg   **(ii)** 75 kg?

    b) Write down an expression for the weight of Toni if Peter weighs q kg.

8 Janet bought a 3 kg bag of potatoes.

    a) How much did the bag cost if 1 kg costs   **(i)** 20 pence   **(ii)** 28 pence?

    b) Write down an expression for the cost of the bag if 1 kg costs t pence.

9

x

x x

x

To make a square Penny uses 4 pieces of wood each x cm long.

Write down an expression for the total length of the 4 pieces of wood.

10 Crisps cost 24 pence a packet.

    a) How much does it cost to buy   **(i)** 2 packets   **(ii)** 4 packets?

    b) Write down an expression for the cost of b packets.

11 Justin bought the same number of apples, oranges and pears.

    a) How many pieces of fruit did he have if the number he bought of each was **(i)** 3   **(ii)** 5?

    b) Write down an expression for the number of pieces of fruit he had if the number he bought of each was h.

12 At Fred's Fishery, chips cost c pence a bag.

    Write down an expression for the cost of   **a)** 2 bags   **b)** 4 bags   **c)** 8 bags.

Number patterns

Look at these dots.

Each column has one more dot than the previous one.
So, if the next column is drawn it would have 5 dots,
and the next one would have 6 dots.

These form a number pattern.
The number of dots in the pattern is 1, 2, 3, 4, 5, 6, ...
We can describe this in words by saying:
'The pattern starts with 1 and goes up 1 each time.'

EXAMPLE 7

Look at the pattern of triangles made up from 1 cm lines.

a) Draw the next pattern.

b) Fill in the missing values in the table.

Pattern number	1	2	3	4	5	6
Number of 1cm lines	3	5				

c) Describe the pattern of 1cm lines in words.

a)

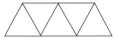

b) 7, 9, 11, 13 in table

c) The pattern starts with 3 and goes up 2 each time.

EXAMPLE 8

Look at this pattern.

15, 13, 11, 9, ...

a) Write down the next two terms of the pattern.

b) Describe the pattern in words.

a) 7, 5

b) The pattern starts with 15 and goes down 2 each time.

EXERCISE 5.3A

1 a) Add the next pattern.

b) Write down the next two terms in the pattern of the number of dots.

c) Describe this pattern in words: 1, 3, 5, ...

2 a) Add the next pattern of black and white squares.

b) Write down the next two terms in the pattern of the number of black squares.

c) Describe this pattern in words: 0, 2, 4, ...

3 a) Add the next pattern of squares.

b) Write down the next two terms in the pattern of the number of squares.

c) Describe this pattern in words: 1, 2, 4, ..., ...

4 For each of the number patterns below:

(i) write down the next two terms in the pattern

(ii) describe the pattern in words.

a) 1, 4, 7, 10, ...

b) 2, 4, 6, 8, ...

c) 20, 19, 18, 17, ...

EXERCISE 5.3B

1 **a)** Add the next pattern of triangles.

  b) Write down the next two terms in the pattern of the number of triangles.

  c) Describe this pattern in words: 1, 3, 5, ..., ...

2 **a)** Add the next pattern of squares.

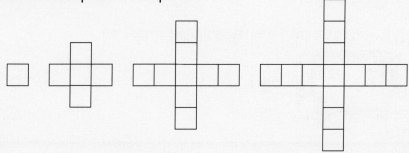

  b) Write down the next two terms in the pattern of the number of squares.

  c) Describe this pattern in words: 1, 5, 9, ..., ...

3 **a)** Add the next pattern of lines.

  b) Write down the next two terms in the pattern of the number of lines.

  c) Describe this pattern in words: 1, 4, 7, ...

4 For each of the number patterns below:

  (i) write down the next two terms in the pattern

  (ii) describe the pattern in words.

  a) 2, 6, 10, 14, ..., ...

  b) 20, 18, 16, 14, ..., ...

  c) 100, 101, 102, 103, ..., ...

Function machines

A function machine consists of an input, a rule and an output.

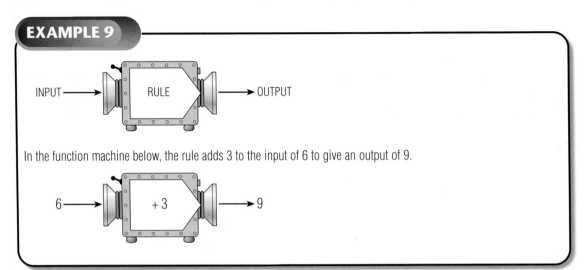

In the function machine below, the rule adds 3 to the input of 6 to give an output of 9.

EXERCISE 5.4A

Find the outputs from each of these function machines.

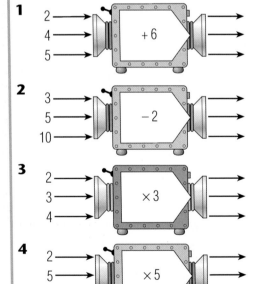

1

2 ⟶
4 ⟶   + 6
5 ⟶

2

3 ⟶
5 ⟶   − 2
10 ⟶

3

2 ⟶
3 ⟶   × 3
4 ⟶

4

2 ⟶
5 ⟶   × 5
8 ⟶

Exercise 5.4A cont'd

5

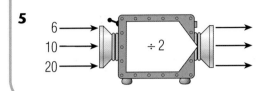

EXERCISE 5.4B

Find the outputs from each of these function machines.

1

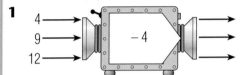

2

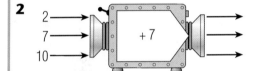

3

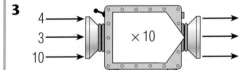

4

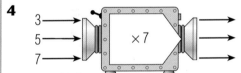

5

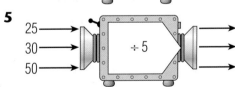

A function machine can have two or more rules in a row.

EXAMPLE 10

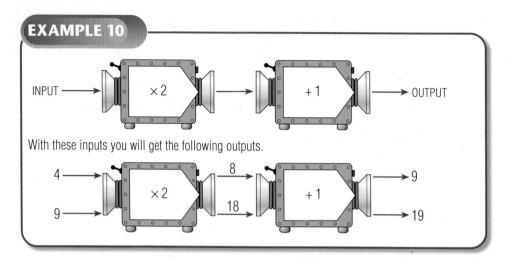

INPUT → ×2 → +1 → OUTPUT

With these inputs you will get the following outputs.

4 → ×2 → 8, 18 → +1 → 9

9 → 19

EXERCISE 5.5A

Find the missing outputs.

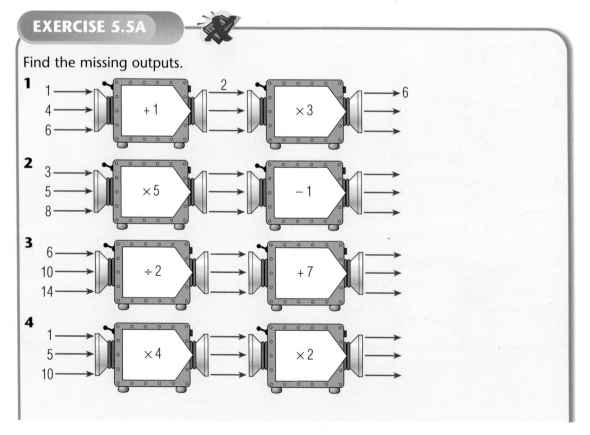

1
1 → +1 → 2 → ×3 → 6
4 →
6 →

2
3 → ×5 → → −1 →
5 →
8 →

3
6 → ÷2 → → +7 →
10 →
14 →

4
1 → ×4 → → ×2 →
5 →
10 →

Chapter 5 *Algebra patterns and using letters for numbers*

Exercise 5.5A cont'd

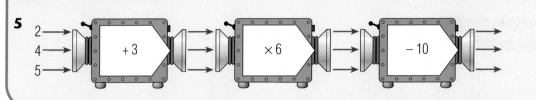

5

Find the missing outputs.

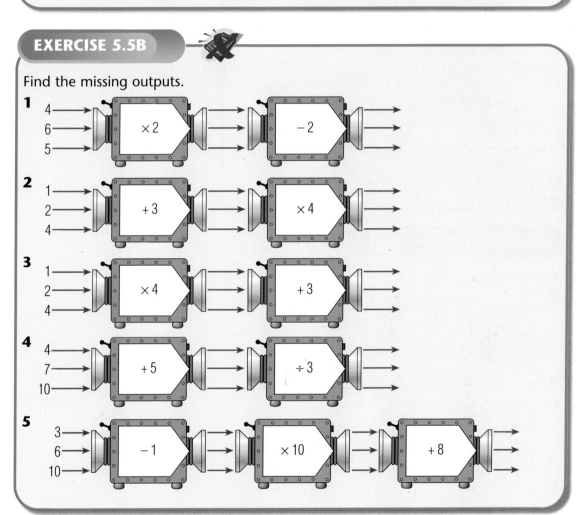

1

2

3

4

5

Chapter 5 *Algebra patterns and using letters for numbers*

Sometimes you will need to find the rule for the function machine.

EXAMPLE 11

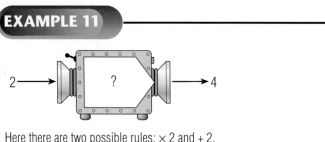

Here there are two possible rules: × 2 and + 2.

Sometimes there is more than one input and output pair. If so, then the rule must work for all of them. So for the function machine below,

EXERCISE 5.6A

Find the missing rules.

1

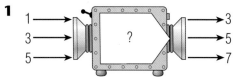

2

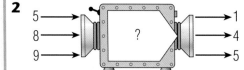

3

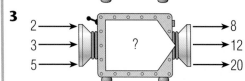

Exercise 5.6A cont'd

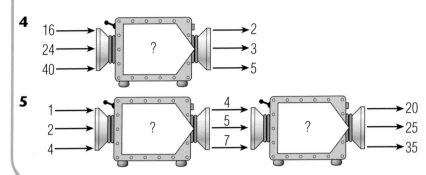

4

16 ⟶ ? ⟶ 2
24 ⟶ ⟶ 3
40 ⟶ ⟶ 5

5

1 ⟶ ? ⟶ 4
2 ⟶ ⟶ 5
4 ⟶ ⟶ 7

? ⟶ 20
⟶ 25
⟶ 35

EXERCISE 5.6B

Find the missing rules.

1

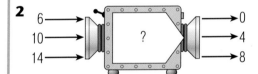

2 ⟶ ? ⟶ 12
4 ⟶ ⟶ 14
5 ⟶ ⟶ 15

2

6 ⟶ ? ⟶ 0
10 ⟶ ⟶ 4
14 ⟶ ⟶ 8

3

3 ⟶ ? ⟶ 18
4 ⟶ ⟶ 24
7 ⟶ ⟶ 42

4

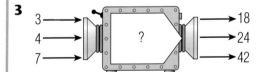

4 ⟶ ? ⟶ 2
8 ⟶ ⟶ 4
12 ⟶ ⟶ 6

5

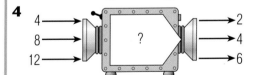
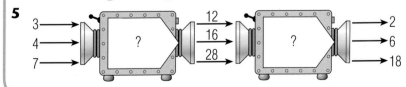

3 ⟶ ? ⟶ 12
4 ⟶ ⟶ 16
7 ⟶ ⟶ 28

? ⟶ 2
⟶ 6
⟶ 18

75

When you know the rule and the output, you can find the input by working backwards and doing the opposite rule.

EXAMPLE 12

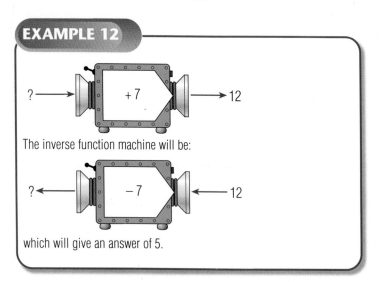

The inverse function machine will be:

which will give an answer of 5.

Exam tip

You can sometimes work these out in your head but it is always worth showing your working out.

EXERCISE 5.7A

Find the missing input values.

1

? → → 8
? → +5 → 10
? → → 13

2

? → → 10
? → −9 → 12
? → → 15

3

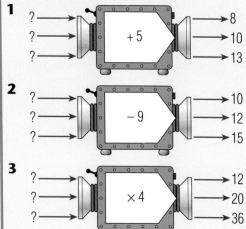

? → → 12
? → ×4 → 20
? → → 36

Exercise 5.7A cont'd

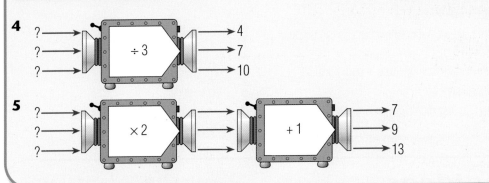

4

? → ÷ 3 → 4
? → → 7
? → → 10

5

? → ×2 → +1 → 7
? → → → 9
? → → → 13

EXERCISE 5.7B

Find the missing input values.

1

? → + 3 → 4
? → → 6
? → → 7

2

? → − 10 → 3
? → → 4
? → → 8

3

? → × 6 → 12
? → → 30
? → → 48

4

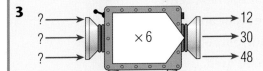

? → ÷ 4 → 3
? → → 5
? → → 10

5

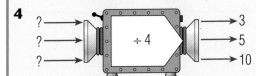

? → + 5 → ÷ 2 → 5
? → → → 6
? → → → 7

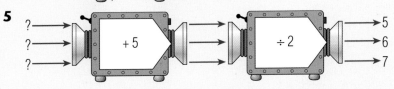

ACTIVITY

Write down as many function machines as you can that start with 2 and end with 6.

Don't forget to use more than one rule in your machine.

Check with your neighbour.

Did you get any that they didn't?

Key ideas

- An expression is a combination of letters and numbers.
- Letters added to pure numbers cannot be combined into a single expression, e.g. 5 plus $a = 5 + a$.
- Letters subtracted from pure numbers cannot be combined into a single expression, e.g. 8 minus $b = 8 - b$.
- Letters and numbers can be multiplied to give a single expression, e.g. k multiplied by $4 = 4k$.
- Letters and numbers can be divided to give a single expression, e.g. m divided by $2 = \frac{m}{2}$.
- Letters which are the same can be combined.
- To describe a sequence you need to state the first term of the sequence together with what you must do to one term to obtain the next term.
- A function machine consists of an input, an output and a rule.
- When finding a rule, it must work for all input/output pairs.
- When finding the input, the inverse rule can be used.

6 Solving problems

In the first pair of exercises you must not use a calculator or any other calculating aid.

EXERCISE 6.1A

1   Don has £69 in his savings account. He adds £25 more. How much does he have now?

2   Sophie has 41 stamps. Her brother has 32 more than her. How many stamps does Sophie's brother have?

3   There are three Year 10 classes in school. There are 24 students in one class, 28 in another and 31 in the third class. How many students are there altogether in the three classes?

4   A box contains 35 chocolates. 8 are eaten. How many are left?

5   A piece of wood is 89 cm long. Anne cuts off 63 cm to finish her bookcase. What length of wood is left?

6   I buy a CD costing £12. I pay for it with a £50 note. How much change will I get?

7   A box of chocolates costs £5. How much will 9 boxes cost?

8   There are 7 rows of chairs in the church hall. Each row contains 8 chairs. How many chairs are there altogether?

9   The school minibus has 14 seats. How many seats are there on 3 such minibuses?

10 A farmer collects 42 eggs from the hens. He puts them into boxes which will each hold 6 eggs. How many boxes will he fill?

11 Three friends win £27 in a raffle. They share the money equally. How much does each receive?

12 A piece of ribbon 1 metre long is cut into pieces 10 cm long. How many pieces will there be?

EXERCISE 6.1B

1   A chocolate bar costs 35p and a packet of crisps costs 28p. How much will they cost altogether?

2   There are 17 people on the bus. At the next stop another 18 get on. How many people are there on the bus now?

3   Rebecca collects £58 for charity. Emma collects £33. How much is this in total?

4   Mum makes 78 sandwiches for a party. 53 are eaten. How many are left?

5   Gran is 80 years old and Elaine is 16. How many years older than Elaine is Gran?

6   Two numbers added together equal 42. One of the numbers is 29. What is the other number?

7   How many days are there in 4 weeks?

8   There are 12 eggs in a box. How many eggs are there in 5 boxes?

9   A lollipop costs 8p. How much will 6 lollipops cost?

10 In a PE lesson 24 students are divided into 3 teams. How many students are there in each team?

11 How many 20p stamps can I buy for £1?

12 A lottery winner gives half of his £38 winnings to charity. How much is this?

EXERCISE 6.2A

In this exercise you may use a calculator if you wish. However, try to answer the question without one first.

1   A shirt costs £33·75 and a tie costs £15·95. How much will they cost altogether?

2   A sack of potatoes weighs 55 kg. How much will 17 of these sacks weigh?

3   There are 684 bars of chocolate to be packed into boxes. 36 chocolate bars are packed into each box. How many boxes will be needed?

4   A holiday costs £489. How much will it cost 5 friends to go on this holiday?

5   A maths book costs £17·95. I pay for it with a £20 note. How much change will I get?

EXERCISE 6.2B

1   A firm makes 2463 Easter eggs on one day and 1895 eggs on the next day. How many Easter eggs have they made altogether?

2   How many 6 m lengths of wire can be cut from a piece 228 m long?

3   Christmas dinner costs £18·70 per person. How much will it cost 4 people to have Christmas dinner?

4   At the beginning of the year a car had driven 10 703 miles. At the end of the year it had driven 19 241 miles. How many miles had it gone during the year?

5   What is the total cost of 16 stamps at 31p each?

Fractions

Fraction means 'part'. So a fraction can tell you how much of a shape is shaded in.

> The bottom number in a fraction is called the **denominator**. This tells you how many equal parts there are.
>
> The top number in a fraction is called the **numerator**. This tells you how many of the equal parts you require.

EXAMPLE 1

What fraction of this shape is   **a)** shaded
                                     b) unshaded?

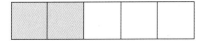

a) Since 2 of the 5 equal parts are shaded, $\frac{2}{5}$ are shaded.

b) Since 3 out of the 5 parts are unshaded, $\frac{3}{5}$ are unshaded.

Exam tip

Remember that to find a fraction of something it must be divided into **equal** parts.

EXERCISE 6.3A

What fraction of the shape is
a) shaded
b) unshaded?

1

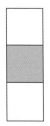

2

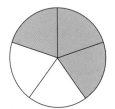

3

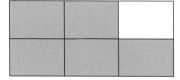

4

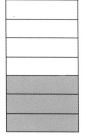

5

EXERCISE 6.3B

What fraction of the shape is
a) shaded
b) unshaded?

1

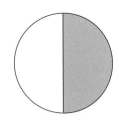

2

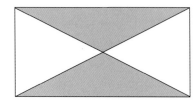

3

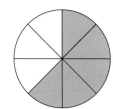

4

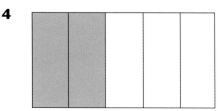

5

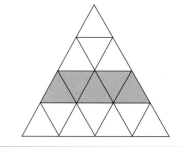

Chapter 6   *Solving problems*

EXERCISE 6.4A

Copy these shapes into your book
Shade the fraction required.

1  $\frac{2}{3}$

2  $\frac{3}{10}$

3  $\frac{5}{8}$

4 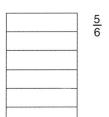 $\frac{5}{6}$

5 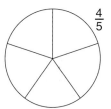 $\frac{4}{5}$

EXERCISE 6.4B

Copy these shapes into your book
Shade the fraction required.

1 $\frac{1}{4}$

2 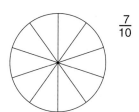 $\frac{7}{10}$

3  $\frac{4}{5}$

4  $\frac{2}{6}$

5 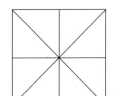 $\frac{3}{8}$

Sometimes the shape may not have the number of equal parts that you want.

EXAMPLE 2

Shade $\frac{4}{5}$ of this shape.

You need 5 equal parts. Since there are 10 squares altogether, grouping them in twos will give 5 equal parts.

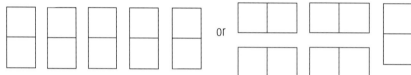

Shade any 4 of these equal parts.

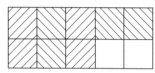

EXERCISE 6.5A

Copy these shapes into your book
Shade the fraction required.

1  $\frac{1}{4}$ **2**  $\frac{3}{10}$ **3**  $\frac{1}{2}$

4 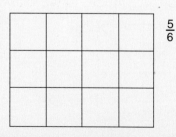 $\frac{5}{6}$ **5** 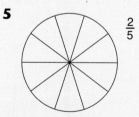 $\frac{2}{5}$

EXERCISE 6.5B

Copy these shapes into your book
Shade the fraction required.

1  $\frac{1}{2}$

2 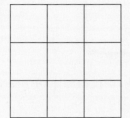 $\frac{2}{3}$

3 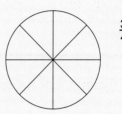 $\frac{3}{4}$

4 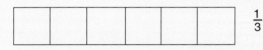 $\frac{1}{3}$

5  $\frac{5}{8}$

Finding $\frac{1}{2}$, $\frac{1}{4}$ or $\frac{3}{4}$ of an amount

When finding $\frac{1}{2}$ of something you divide it by 2. That is, you split it into two equal parts.

EXAMPLE 3

a) Find $\frac{1}{2}$ of 10.

Answer = 5.   [10 ÷ 2 = 5 and 5 + 5 = 10]

b) Find $\frac{1}{2}$ of 3.

Answer = 1·5 or $1\frac{1}{2}$ [3 ÷ 2 = 1·5 and 1·5 + 1·5 = 3]

When finding $\frac{1}{4}$ of something you divide it by 4. To do this you can halve it and halve it again.

EXAMPLE 4

a) Find $\frac{1}{4}$ of 20.

Answer = 5. $[\frac{1}{2}$ of 20 = 10 and $\frac{1}{2}$ of 10 = 5]

b) Find $\frac{1}{4}$ of 6.

Answer = 1·5 $[\frac{1}{2}$ of 6 = 3 and $\frac{1}{2}$ of 3 = 1·5]

When finding $\frac{3}{4}$ of something, find $\frac{1}{2}$ of it and $\frac{1}{4}$ of it and add the two answers together.

EXAMPLE 5

Find $\frac{3}{4}$ of 80.

$\frac{1}{2}$ of 80 = 40 $\frac{1}{4}$ of 80 = 20 $[\frac{1}{2}$ of 80 = 40 and $\frac{1}{2}$ of 40 = 20]

Answer = 40 + 20 = 60.

EXERCISE 6.6A

1 Find $\frac{1}{2}$ of the following.

   a) 100   **b)** 30   **c)** 12

   d) 15   **e)** 27

2 Find $\frac{1}{4}$ of the following.

   a) 24   **b)** 80   **c)** 16

   d) 18   **e)** 50

3 Find $\frac{3}{4}$ of the following.

   a) 20   **b)** 60   **c)** 16

   d) 10   **e)** 14

EXERCISE 6.6B

1 Find $\frac{1}{2}$ of the following.

   a) 20   **b)** 16   **c)** 70

   d) 13   **e)** 25

2 Find $\frac{1}{4}$ of the following.

   a) 40   **b)** 28   **c)** 100

   d) 22   **e)** 30

3 Find $\frac{3}{4}$ of the following.

   a) 6   **b)** 12   **c)** 80

   d) 18   **e)** 30

ACTIVITY

See if you can solve the following problems.

a) $\frac{1}{2}$ of $?= 25$ **b)** $\frac{1}{4}$ of $?= 14$ **c)** $\frac{3}{4}$ of $?= 45$

Can you explain to your neighbour how you found your answer?
Draw a poster to explain your method.

Key ideas

- A question can be asking you to add, subtract, multiply or divide.
- You should know how to add, subtract, multiply or divide whole numbers or money without a calculator.
- You should be able to use a calculator to solve problems.
- You should be able to interpret the display on a calculator.
- An answer to a money problem needs two numbers after the decimal point.
- The parts of a fraction are called the 'numerator' and the 'denominator'.
- You should know what the 'numerator' tells you about the fraction.
- You should know what the 'denominator' tells you about the fraction.
- A fraction involves equal parts.
- $\frac{1}{4}$ of something can be found by halving twice.
- $\frac{3}{4}$ of something can be found by adding together $\frac{1}{2}$ and $\frac{1}{4}$ of that amount.

C1 Revision exercise

1 Write expressions for the lengths of the lines below.

a)

b)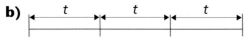

c)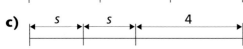

2 Write expressions for the missing lengths (marked *?*) on the lines below.

a)

b)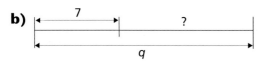

c)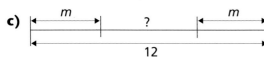

3 Apples cost *y* pence a pound. Write down expressions for the cost of:

a) 2 pounds   **b)** 5 pounds
c) 7·5 pounds.

4 Tom is 25 years older than his son.

a) How old will he be when his son's age is:
   (i) 5   **(ii)** 12?

b) Write down an expression for Tom's age when his son is *a* years old.

5 A packet of Spenders crisps is *c* pence dearer than a packet of Givers crisps.

Write down an expression for the cost of a packet of Spenders crisps when a packet of Givers crisps costs:

a) 20 pence   **b)** 27 pence.

6

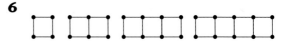

a) Add the next pattern to the diagram.

b) Complete this table for the diagram.

Pattern	1	2	3	4	5	6
Number of dots	4	6				
Number of lines	4	7				

c) Describe the pattern of:
   (i) the number of dots
   (ii) the number of lines.

7 For each of the number patterns below

a) add two more terms

b) describe the pattern in words.

   (i) 1, 3, 5, 7, ... , ...
   (ii) 1, 5, 25, 125, ... , ...
   (iii) 25, 22, 19, 16, ... , ...

8 Complete the following function machines.

a)

b)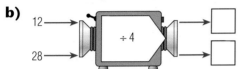

c)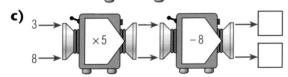

d)

e)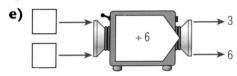

9

a) Beverley saves £3·50 a week for 6 weeks. How much does she save altogether?

b) There are 41 people on the bus. At the next stop 16 get off and 38 get on. How many people are there on the bus now?

c) A magazine costs £1·85. If I am given 65p change, how much money did I give the newsagent?

d) 5 people share £120 that they won on the National Lottery. How much do they each get?

e) At the Steelers' last three games the attendances were 5831, 2387 and 1579.
How many people attended all three games?

10 a) What fraction of the circle is
(i) shaded **(ii)** unshaded?

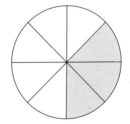

b) Copy this shape. Shade in $\frac{4}{7}$ of it.

11 Copy this shape.
Shade in $\frac{1}{4}$ of it.

12 a) Find $\frac{1}{2}$ of
(i) 16
(ii) 31
(iii) 164

b) Find $\frac{1}{4}$ of
(i) 60
(ii) 26
(iii) 180

c) Find $\frac{3}{4}$ of
(i) 36
(ii) 84
(iii) 50

7 Shapes

Triangles

These shapes are all triangles. How can you sort them into different groups?

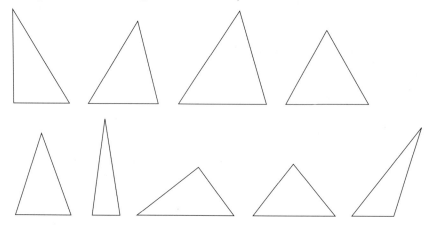

These are **right-angled triangles**.

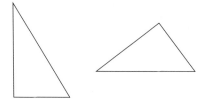

These are **isosceles triangles**.
They have two sides the same.

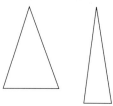

You can show the equal sides by
marking them with a line like this.

These are **equilateral triangles**.
All their sides are the same.
Ali their angles are the same.

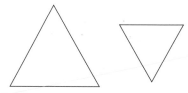

The triangles on the right are **scalene triangles**.
In scalene triangles all the sides and all the angles are different.

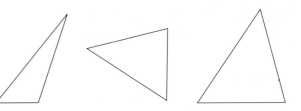

Quadrilaterals

Four-sided shapes are called **quadrilaterals**.
You know the **square** and **rectangle**.
Here are some other special quadrilaterals.

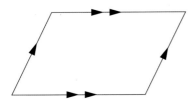

This is a **parallelogram**.
It has two pairs of parallel sides.

The quadrilateral on the right is a **rhombus**.
It is a parallelogram, but, also, all its sides are the same length.

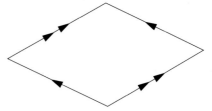

This is a **kite**.
It has two pairs of equal sides.
The equal sides are next to each other, not opposite as they are in a parallelogram or rectangle.

This is a **trapezium**.
It has one pair of parallel sides.

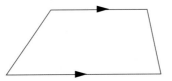

These shapes are not as common as squares and rectangles, but they are used a great deal. Can you see any examples of them in the room around you?

Polygons

A **polygon** is a many-sided shape.

Here are some common ones, apart from the triangles and quadrilaterals you have met already.

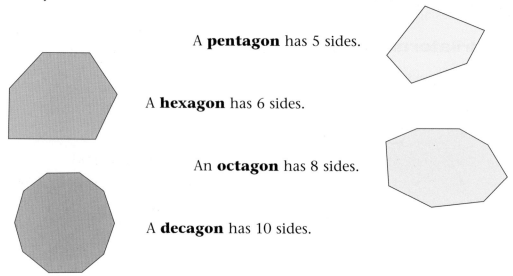

A **pentagon** has 5 sides.

A **hexagon** has 6 sides.

An **octagon** has 8 sides.

A **decagon** has 10 sides.

When the sides and angles of a polygon are all the same, it is called a **regular polygon**.

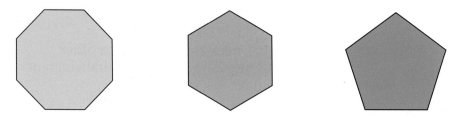

Constructing a regular polygon

You will need a pair of compasses and a protractor or angle measurer.

EXAMPLE

Construct a regular hexagon (6 sides).

Before you can draw any of the regular polygons you need to find the angle at the centre of the shape.

There are 6 equal angles at the centre of a regular hexagon.

Since one complete turn is 360°, each of these angles will be $\frac{360}{6} = 60°$.

Now you are ready to draw the shape.

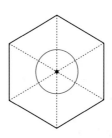

1   Draw a circle with a radius of 4 cm.

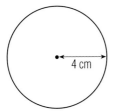

4 cm

2   Mark a point at the top of the circle.

Measure an angle of 60° at the centre and mark a second point.

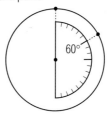

60°

3   Choose method **a)** or **b)**.

a) Continue measuring angles of 60° and mark points all around the circle.   **or**

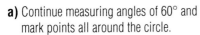

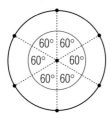

60° 60°
60°   60°
60° 60°

b) Open your compass to the gap between the two points. Mark off this gap around the circle until you get back to the top.

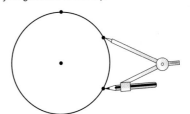

Example cont'd

4 Join the points to make the regular hexagon.
You could rub out the circle if you wanted to.

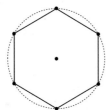

1 Construct a regular pentagon (5 sides).

2 Construct a regular octagon (8 sides).

3 Construct a regular decagon (10 sides).

4 Construct a regular quadrilateral (4 sides – a square).

5 Construct a regular triangle (3 sides – an equilateral triangle).

6 Construct a regular nonagon (9 sides).

ACTIVITY 1

Construct a square.

Join the middle points of each side to form a smaller square inside the first square.

Continue to do this to make a pattern.

Colour in your pattern.

Make a poster.

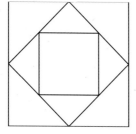

ACTIVITY – EXTENSION

Draw the same type of pattern for a pentagon or an octagon, etc.

ACTIVITY 2

Mark the points for a regular hexagon around the circle. Join them to make this shape.

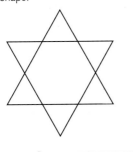

ACTIVITY – EXTENSION

Try to make other shapes by using the points for other regular polygons.

Enlargements

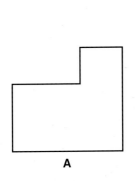

A

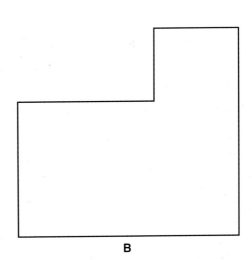

B

Every length in diagram B is 2 times as long as those in diagram A.

We say that diagram B is an **enlargement** of diagram A.

The number of times one diagram is bigger than another is called the **scale factor** of the enlargement. Here the scale factor is 2.

EXERCISE 7.2A

1 Draw enlargements of the following diagrams. Use the scale factor given.

a)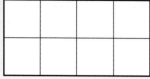

Scale factor 2

b)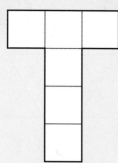

Scale factor 3

c)

Scale factor 2

d)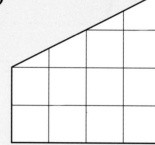

Scale factor 3.

e) Enlarge the first letter of your first name and the first letter of your last name by a scale factor of 2. Do this on centimetre-square paper.

Exercise 7.2A cont'd

2 Measure the lengths in the two diagrams to see if one is an enlargement of the other. If it is, give the scale factor of the enlargement. If it is not, say why.

a)

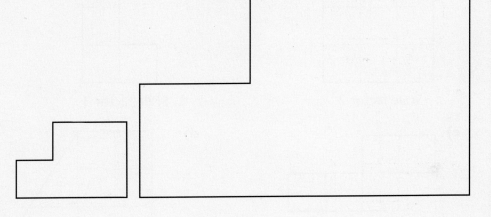

b)

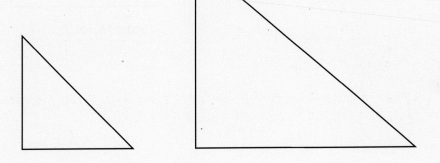

EXERCISE 7.2B

1 Draw enlargements of the following diagrams. Use the scale factor given.

a)

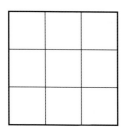

Scale factor 2

b)

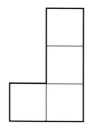

Scale factor 3

c)

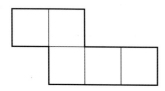

Scale factor 3

d)

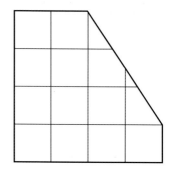

Scale factor 2

Exercise 7.2B cont'd

2 Measure the lengths in the two diagrams to see if one is an enlargement of the other. If it is, give the scale factor of the enlargement. If it is not, say why.

a)

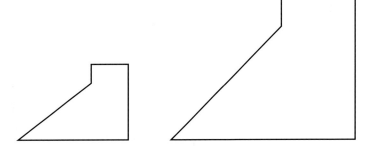

b)

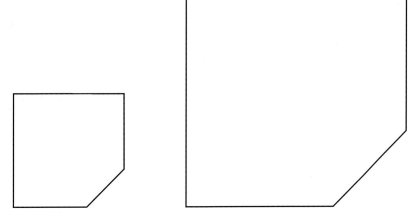

ACTIVITY 3

On centimetre-square paper draw an everyday shape like the one below.

Enlarge it by a scale factor of your own choice.

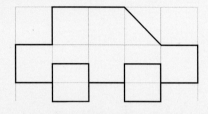

Key ideas

- You should know the names and properties of special triangles and quadrilaterals.
- A polygon is a many-sided shape.
- You should know the number of sides and names of the more common polygons.
- A regular polygon has sides of equal length and equal angles.
- You should know how to construct regular polygons using compasses and protractor.
- The scale factor of an enlargement tells you how many times bigger the shape needs to be drawn.
- The enlargement means every length is increased by the same scale factor.

8 Units, perimeter, area and volume

Lengths

Lengths can be measured
either in imperial (old) units – inches, feet, yards, miles
or in metric units–millimetres, centimetres, metres, kilometres.
You may have used feet and
inches to measure your height,

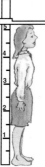

Height = 5 foot 3 inches.

and miles to measure distance.

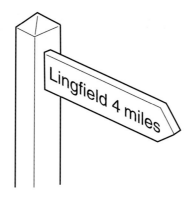

You may not have used yards at all as they are not used very often nowadays.

Much more important nowadays is the metric system.

You may use millimetres or centimetres to measure your handspan, metres or centimetres to measure your height,

Height = 1·60 m or 160 cm.

Handspan = 190 mm or 19 cm.

and kilometres to measure a long distance.

You do not need to write out the words in full each time. The accepted short versions are metre = m, millimetre = mm, centimetre = cm, kilometre = km.

EXAMPLE 1

What metric measure would you use to give
a) the width of a desk
b) the length of the school playing field
c) the distance from Birmingham to London?

a) millimetres or centimetres
b) metres
c) kilometres.

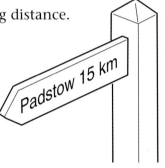

Exam tip

Most short lengths can be given in either centimetres or millimetres.

ACTIVITY 1

Look at these numbers.

21·375   145   0·003 24   64·8   7·0132   789·2

Use a calculator to work out each of the following calculations.

1   Multiply each of the numbers by 10. What do you notice?

2   Multiply each of the numbers by 100. What do you notice?

3   Multiply each of the numbers by 1000. What do you notice?

4   Divide each of the numbers by 10. What do you notice?

5   Divide each of the numbers by 100. What do you notice?

6   Divide each of the numbers by 1000. What do you notice?

Try to write down a summary of what happens in each case.

The connections between these units are

1 metre = 1000 millimetres	1m = 1000mm
1 metre = 100 centimetres	1m = 100cm
1 centimetre = 10 millimetres	1cm = 10mm

These are important and you need to be able to change between them.

Because the connections are all multiples of 10, changing from one to another never changes the figures, it only changes the position of the decimal point or adds or deletes zeros.

Exam tip

When changing from one metric unit to another, decide what to multiply or divide by and then move the decimal point by that number of places. Whole numbers can have a decimal point after the number and when dividing it is helpful to put in the decimal point.

EXAMPLE 2

Change these lengths in metres to millimetres

a) 2·435 m   **b)** 3·52 m   **c)** 4 m   **d)** 3·05 m.

As there are 1000 millimetres in a metre, to change them you just multiply by 1000.

This means that the decimal point needs to be moved 3 places to the right or, if the number is a whole number, 3 zeros must be added.

a) 2·435 × 1000 = 2435 mm   The point has moved 3 places.

b) 3·52 × 1000 = 3520 mm   To move the point 3 places it is necessary to add a zero.

c) 4 × 1000 = 4000 mm   A whole number so add 3 zeros.

d) 3·05 × 1000 = 3050 mm   Again 1 extra zero is needed.

EXAMPLE 3

Change these lengths to millimetres.

a) 15 cm   **b)** 1·5 cm   **c)** 4·625 m

a) 15 × 10 = 150 mm   1 cm = 10 mm, so times by 10, a whole number, so add a zero.

b) 1·5 × 10 = 15 mm   As in **a)** but this time move the decimal point one place.

c) 4·625 × 1000 = 4625 mm   1 m = 1000 mm so times by 1000.

EXAMPLE 4

Change these lengths to centimetres.

a) 5·25 m   **b)** 2·542 m   **c)** 20 mm   **d)** 57 mm   **e)** 42·5 mm

a) 5·25 × 100 = 525 cm   1 m = 100 cm, so times by 100, which moves the decimal point 2 places to the right.

b) 2·542 × 100 = 254·2 cm   The same as in part **a)**.

c) 20 ÷ 10 = 2 cm   1 cm = 10 mm, so divide by 10, which deletes a zero.

d) 57 ÷ 10 = 5·7 cm   The same as in **c)** but start with the decimal point after the whole number and move it one place to the left.

e) 42·5 ÷ 10 = 4·25 cm   The same as in **d)**.

EXAMPLE 5

Change these lengths to metres.

a) 148 cm   **b)** 291·4 cm   **c)** 3360 mm

a) 148 ÷ 100 = 1.48 m   1 m = 100 cm, so divide by 100, which moves the point 2 places to the left.

b) 291·4 ÷ 100 = 2·914 m   The same as **c)**.

c) 3360 ÷ 1000 = 3·36 m   1 m = 1000 mm, so divide by 1000. Move the point 3 places to the left. The zero at the end can be left off after the decimal point.

EXAMPLE 6

Put these lengths in order, smallest first.

3·25 m   415 cm   302 mm   5012·5 mm   62·3 cm

To put these in order, first change them all to the same unit. Normally, it is easiest to change to the smallest unit which, in this case, is mm.

3·25 m   = 3·25 × 1000 mm   = 3250 mm   Multiply by 1000 to change m to mm.

415 cm   = 415 × 10 mm   = 4150 mm   Multiply by 10 to change cm to mm.

302 mm   = 302 mm

5012·5 mm   = 5012·5 mm

62·3 cm   = 62·3 × 10 mm   = 623 mm.

So order is 302 mm, 623 mm, 3250 mm, 4150 mm, 5012·5 mm

or   302 mm, 62·3 cm, 3·25 m, 415 cm, 5012·5 mm.

EXERCISE 8.1A

1   What metric units would you use to give
   a) the width of a book   **b)** the height of a room
   c) the width of an envelope   **d)** the distance from London to Glasgow
   e) the distance round a running track   **f)** the length of a classroom
   g) the length of a finger   **h)** the length of a motor car race?

2   Change these lengths to millimetres.
   a) 4 cm   **b)** 33 cm   **c)** 2·5 cm   **d)** 52 cm   **e)** 4·52 cm.

Exercise 8.1A cont'd

3 Change these lengths to millimetres.

    a) 9 m   **b)** 1·129 m   **c)** 3·1 m   **d)** 0·3 m   **e)** 2·101 m.

4 Change these lengths to centimetres.

    a) 4 m   **b)** 5·22 m   **c)** 9·16 m   **d)** 8·275 m   **e)** 52 m.

5 Change these lengths to centimetres.

    a) 20 mm   **b)** 140 mm   **c)** 35 mm   **d)** 94·6 mm   **e)** 660 mm.

6 Change these lengths to metres.

    a) 142 cm   **b)** 4570 cm   **c)** 9124 mm   **d)** 5800 mm   **e)** 2146·3 mm.

7 Write these lengths in order of size, smallest first.

    2·42 m   1623 mm   284 cm   9·044 m   31·04 cm.

EXERCISE 8.1B

1 What metric units would you use to give

    a) the length of a swimming pool   **b)** the height of a church steeple

    c) the length of a nail   **d)** the width of a window

    e) the distance round your waist   **f)** the length of a cross-country race?

2 Change the following lengths to millimetres.

    a) 2 cm   **b)** 4·5 cm   **c)** 9·35 cm   **d)** 219 cm   **e)** 99·1 cm.

3 Change these lengths to millimetres.

    a) 3 m   **b)** 2·239 m   **c)** 9·1 m   **d)** 4·3 m   **e)** 0·124 m.

4 Change these lengths to centimetres.

    a) 5 m   **b)** 2·32 m   **c)** 18·16 m   **d)** 3·295 m   **e)** 4·1952 m.

5 Change these lengths to centimetres.

    a) 70 mm   **b)** 310 mm   **c)** 46 mm   **d)** 8000 mm   **e)** 1480 mm.

6 Change these lengths to metres.

    a) 5142 mm   **b)** 570 cm   **c)** 1146 mm   **d)** 580·4 cm   **e)** 41 623 mm.

7 Write these lengths in order of size, smallest first.

    423 cm   6123 mm   804 cm   3·211 m   105 mm.

Perimeter, area and volume

Perimeter

The perimeter of a shape is the distance all the way around the edge of the shape.

Since the perimeter of a shape is a length, you must use units such as centimetres (cm), metres (m), and kilometres (km).

Simple perimeters

EXAMPLE 7

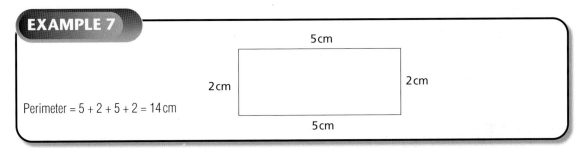

Perimeter = 5 + 2 + 5 + 2 = 14 cm

Shapes can be irregular, but you work out the perimeter in exactly the same way.

EXAMPLE 8

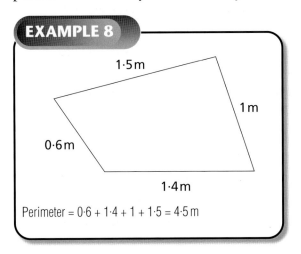

Perimeter = 0·6 + 1·4 + 1 + 1·5 = 4·5 m

Exam tip

Though you don't have to give the units in your working, you must remember to give the units with your answer.

EXERCISE 8.2A

Find the perimeter of the following shapes.

1

3 cm

8 cm   8 cm

3 cm

2

14 cm

5 cm

3

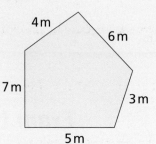

4 m

6 m

7 m

3 m

5 m

4

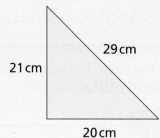

29 cm

21 cm

20 cm

5 A square of side 25 cm.

EXERCISE 8.2B

Find the perimeter of the following shapes.

1

3·5 cm

1·5 cm   1·5 cm

3·5 cm

2

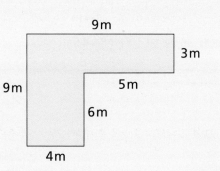

9 m

3 m

9 m   5 m

6 m

4 m

Exercise 8.2B cont'd

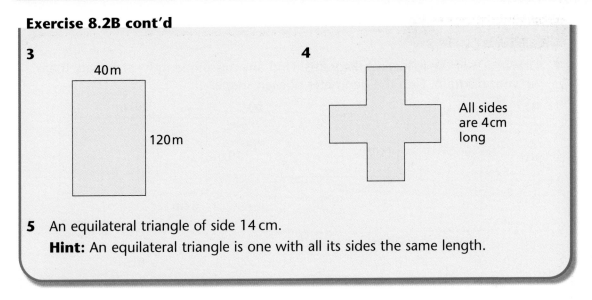

3 40 m / 120 m

4 All sides are 4 cm long

5 An equilateral triangle of side 14 cm.

Hint: An equilateral triangle is one with all its sides the same length.

More complicated perimeters

Sometimes not all of the lengths of the shape are given in the diagram. Before trying to find the perimeter, all lengths must be found.

EXAMPLE 9

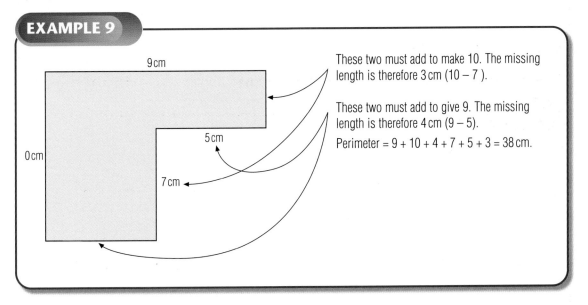

9 cm / 0 cm / 5 cm / 7 cm

These two must add to make 10. The missing length is therefore 3 cm (10 – 7).

These two must add to give 9. The missing length is therefore 4 cm (9 – 5).

Perimeter = 9 + 10 + 4 + 7 + 5 + 3 = 38 cm.

EXERCISE 8.3A

1 Draw each of the following diagrams. Find any missing lengths and mark them on your diagram. Find the perimeter of each shape.

a)

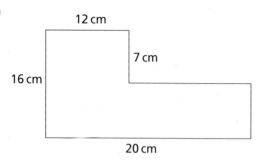

b)

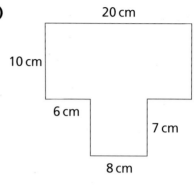

c)

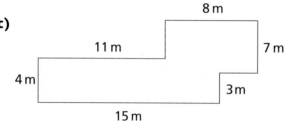

2 Measure the following shapes accurately. Work out the perimeter of each of them.

a)

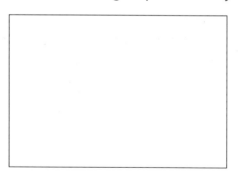

b)

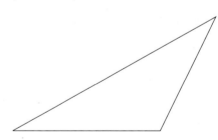

3 The perimeter of a rectangle is 26 cm. One side is 8 cm long. How long is the other side?

Hint: Draw a diagram and label as many lengths as possible.

EXERCISE 8.3B

1 Draw each of the following diagrams. Find any missing lengths and mark them on your diagram. Find the perimeter of each shape.

a)

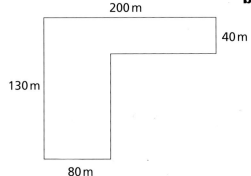

b)

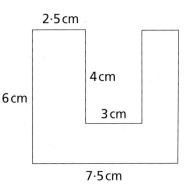

c)

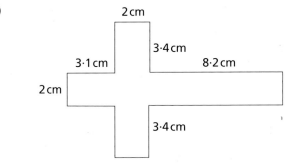

2 Measure the following shapes accurately. Work out the perimeter of each of them.

a)

b)

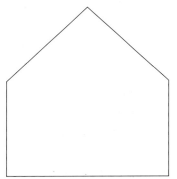

3 A square has a perimeter of 120 cm. How long is each side?

Area

The area of a shape is the amount of space inside the shape.

When you find the area of small shapes you measure the area in square centimetres. This is usually written as cm^2.

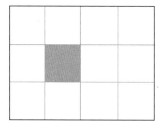

The grid is drawn in centimetres.
The shaded area is a one centimetre square.
Its area is 1 cm^2.

In the same way, a square with sides of 1 kilometre will have an area of 1 km^2.

EXERCISE 8.4A

Find the area of each of the shapes by counting squares. Write your answers in square centimetres.

1 a)

b)

c)

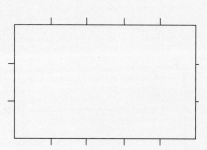

d)

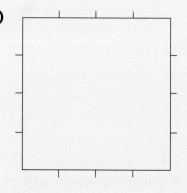

Exercise 8.4A cont'd

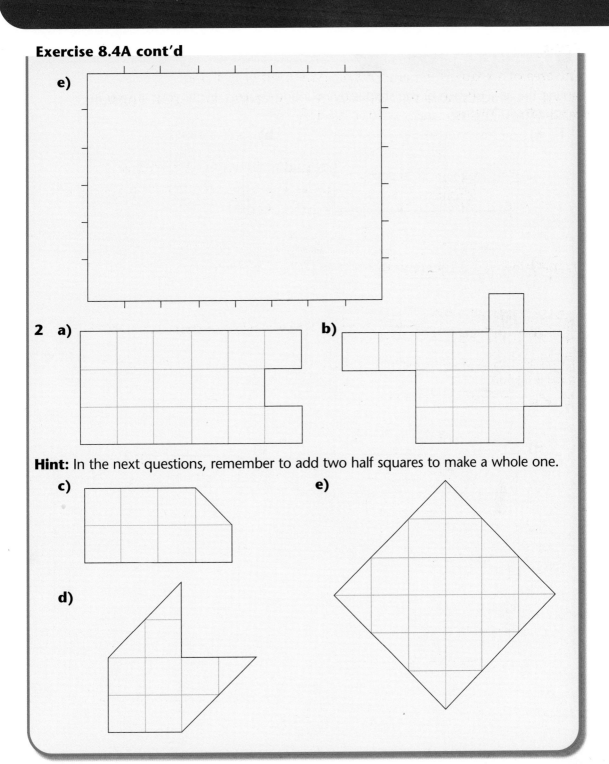

e)

2 a)

b)

Hint: In the next questions, remember to add two half squares to make a whole one.

c)

e)

d)

Chapter 8   *Units, perimeter, area and volume*

EXERCISE 8.4B

Find the area of each of the shapes by counting squares. Write your answer in square centimetres.

1 a)

b)

c)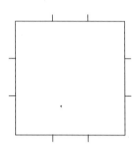

d)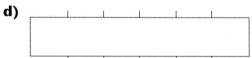

e)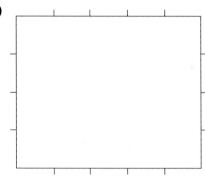

Exercise 8.4B cont'd

2 a)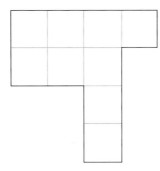

b)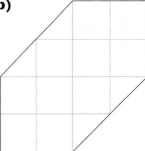

c)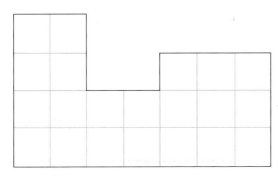

d)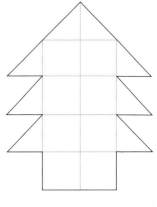

e)

ACTIVITY 2

1 How many different shapes can you draw with a perimeter of 18 cm.

2 How many different rectangles can you draw with a perimeter of 20 cm. Which rectangle has the largest area?

Estimating the area of irregular shapes

Often, shapes cover irregular areas. Though it is difficult to find the area exactly, you can find an estimate of the area it covers.

EXAMPLE 10

Find the approximate area of the shape below.

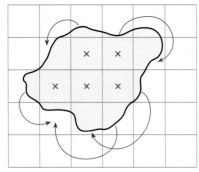

Hint: Put a cross in each square as you count them. You will find that placing a piece of tracing paper over the picture will save you from marking the book.

a) Count how many full squares are covered.

b) 'Piece together' the part squares to make full ones.

Using this method you will find that the area of the shape above is 5 full squares

plus   4 'pieced together' squares

so   the area is approximately $9\,\text{cm}^2$.

EXERCISE 8.5A

1 Estimate the area of each of the following shapes.

a)

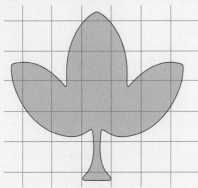

b)

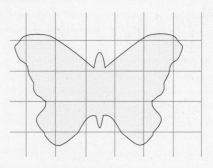

c)

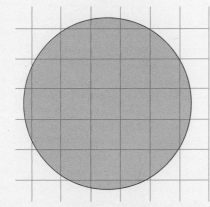

Below is the map of an island. Each square has a side of 1 km. Find an estimate of the area of the island in square kilometres.

d)

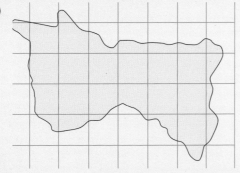

2 On a piece of centimetre-square paper, draw around your hand. Use this drawing to estimate the area of your hand print.

EXERCISE 8.5B

1 Estimate the area of each of the following shapes.

a)

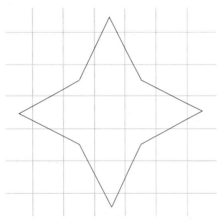

b)

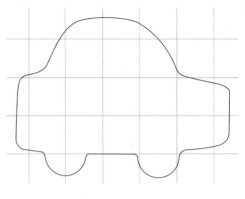

c)

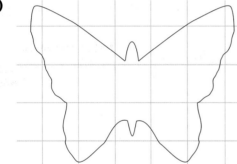

Below is the map of an island. Each square has a side of 1 km. Find an estimate of the area of the island in square kilometres.

d)

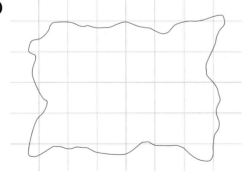

2 On a piece of centimetre-square paper, draw around your foot. Use this to estimate the area of your foot print.

On a piece of centimetre-square paper draw around some of the things in your school bag.

For example
- calculator
- pencil case
- exercise book.

Work out the approximate area of each item,

Volume

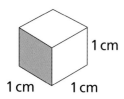

The volume of a three-dimensional shape is the amount of space it takes up.

A cube 1 cm long, 1 cm wide and 1 cm high will have a volume of one cubic centimetre – **1 cm³**.

In the same way cubes with sides of 1 m will have a volume of one cubic metre – **1 m³**.

EXAMPLE 11

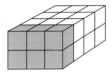

Find the volume of this solid shape.

One layer has 6 cubes. There are 4 layers.

The volume is $6 \times 4 = 24 \, \text{cm}^3$

EXAMPLE 11

Find the volume of this solid shape.

One layer has 4 cubes. There are 3 layers.

The volume is $4 \times 3 = 12 \, \text{cm}^3$

EXERCISE 8.6A

1   In the following diagrams, each small cube is 1 cm long, 1 cm wide and 1 cm
    high. Find the volume of each of the solid shapes.

a)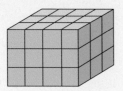

b)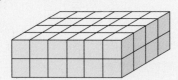

c)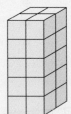

d)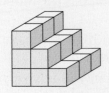

e)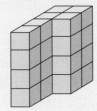

2   How many cubes of volume 1 m³ can fit into a rectanglular box measuring 2 m
    by 3 m by 5 m?

Note: The **capacity** of the box is how much volume is contained within it.

EXERCISE 8.6B

1   In the following diagrams each small cube has a volume of 1 cm³. Find the
    volume of each of the solid shapes.

a)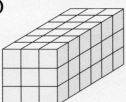

b)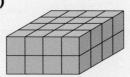

c)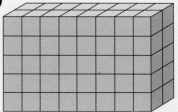

Exercise 8.6B cont'd

d)

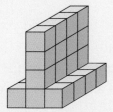

e)

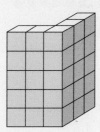

2 How many cubes, each of volume 1 m³, can fit into a rectangular box measuring 4 m by 4 m by 4 m?

Key ideas

- You should know the notation for metres, centimetres and millimetres.
- You should know the connection between metres, centimetres and millimetres and how to convert between them.
- The perimeter is the distance all the way around the edge of a shape.
- The area is the space inside a shape.
- You can only find an estimate of the area inside an irregular shape.
- The volume is the amount of space that a shape occupies.
- You should know to include the appropriate units in an answer when finding perimeter, area or volume, e.g. cm, cm², cm³.

Revision exercise

1 Construct a regular hexagon (6 sides).

2 Draw enlargements of the following diagrams on centimetre-square paper. Use the scale factor given.

   a) Scale factor of 3.

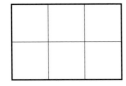

   b) Scale factor of 2.

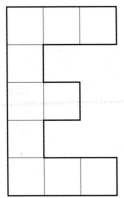

3 John says that Shape B is an enlargement of Shape A. Is he correct? Explain your answer.

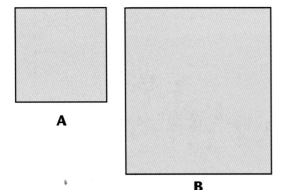

4 What metric units would you use for:

   a) the length of a pencil

   b) the height of a church.

5 Write these amounts in the units indicated.

   a) 3·2 m in mm   **b)** 4·5 m in cm

   c) 15 cm in mm   **d)** 584·2 cm in m

   e) 14 523 mm in m.

6 Put the amounts in order of size, smallest first.

   5 m, 500 mm, 655 cm, 7124 mm, 2·375 m.

7 Estimate the following using metric units:

   a) the length of a swimming pool

   b) the length of a chair leg.

8 Find the perimeters of these shapes.

a)

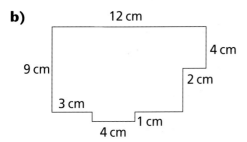

b)

c)

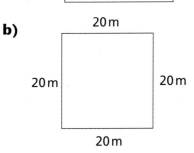

9 Find the area and perimeter of each of these rectangles.

a)

7 cm

3 cm

b)

20 m

20 m   20 m

20 m

10 A piece of A4 paper measures 21 cm by 29·7 cm. Calculate the perimeter of the piece of paper.

11 Find an estimate of the area of the shape below.

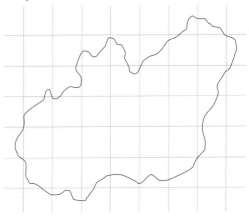

12 In the shapes below, each of the small cubes has a volume of 1 cm³. Find the volume of each solid shape.

a)

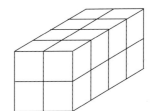

b)

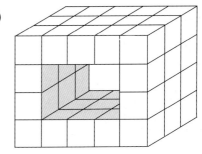

9 Representing data

Pictograms

Pictograms are graphs where the symbol represents a group of units.
The symbols are all the same size and are placed in rows and columns.

EXAMPLE

This is an example of a pictogram showing people's favourite colour.

This symbol represents 2 people

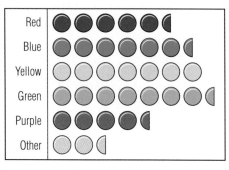

a) What does represent?

b) How many people chose blue?

c) How many people altogether were asked about their favourite colours?

a) 1 person **b)** 13 **c)** 67

EXERCISE 9.1A

1 Emma drew this pictogram to show the number of books borrowed from the school library one week.

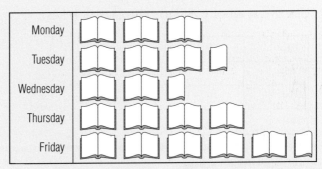

represents 10 books.

a) How many students borrowed books on each of the days.

b) Which day was most popular? Why do you think that might be?

2 Draw a pictogram to show the eye colours of a group of children in school.

Use this symbol   to represent 2 students.

The eye colours are:

blue	18	
blue/green	6	
grey	5	
brown	11	

3 The table below shows the number of CDs sold in a music shop in the week before Christmas:

 represents 20 CDs.

a) What numbers should go in the total column?

b) What were the total sales?

Exercise 9.1A cont'd

4 The table below shows the number of bikes sold by 'Bikes 'R' Us' in four weeks. Draw a pictogram to represent this information.

Use the symbol 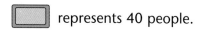 to represent 8 bikes.

Week 1	12
Week 2	10
Week 3	19
Week 4	13

EXERCISE 9.1B

1 The pictogram below shows how many people in a small village watch each of the different TV channels.

represents 40 people.

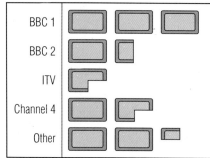

a) Which channel is most popular?

b) Which channel is least popular?

c) Work out how many people watch each of the channels.

2 Geta is doing a survey on the number of letters in a person's first name.

The table shows the data she collected.

Numbers of letters	Number of students
4	6
5	5
6	9
7	2
8	4
9	1

Exercise 9.1B cont'd

a) Draw a pictogram for this information.

Use the symbol   to represent 2 students.

b) Which is the most popular length of first name?

c) Conduct a survey of the number of letters in the first name of the students in your class. Draw a pictogram to show the information. Use a different symbol.

3 The table shows the number of students absent from school one week.

Use these symbols to draw a pictogram to represent the information.

Monday	8
Tuesday	5
Wednesday	7
Thursday	10
Friday	6

4 students   3 students   2 students   1 student

4 Design a data collection sheet to help you produce a pictogram to show which pets students in your class or group own.

Use a complete symbol for 2 people and half a symbol for 1 person. Design your own symbol.

ACTIVITY

Conduct a survey of the students in your class.

Ask everyone something like
- 'Are you left or right handed?'
- 'What month were you born in?'
- 'How many people are in your family?'
- A question of your own.

Draw a pictogram to show your data.

Reading from graphs

Emily is in hospital. Every 3 hours her temperature is taken and the points plotted on a graph.

Time	00:00	03:00	06:00	09:00	12:00	15:00	18:00
Temperature °C	38	38·5	38·2	39·2	39·2	38·6	39·4

The points are joined by straight lines – but does this mean anything? Would anyone be able to state that, at 10:00, Emily's temperature was, say, 39·2 °C? Emily's temperature might vary like the curved line on the graph. The straight lines are drawn to give an impression. The intermediate values don't really have any meaning.

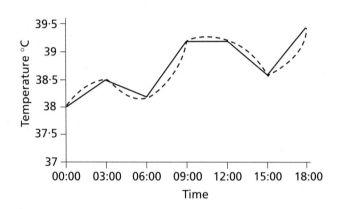

1 This graph shows the average monthly rainfall for a town in Britain measured over a 30-year period.

a) What was the least rainfall?

b) What was the range?

c) Which two months had an average rainfall of 70 mm?

d) Are you able to say with certainty what the rainfall is in the middle of each month?

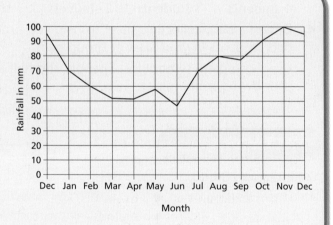

Exercise 9.2A cont'd

2 Peggy is the caretaker at a school.

She makes sure the heating in the school is kept at a suitable temperature. She takes the temperature several times each day and draws graphs like this.

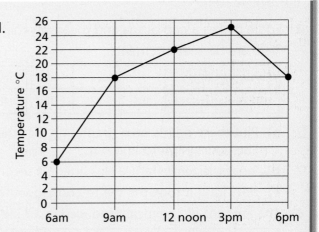

a) What was the highest temperature?

b) What was the temperature at **(i)** 10 am **(ii)** 5 pm?

3 The sales in a local shop over the course of a week are given below.

Day	Monday	Tuesday	Wednesday	Thursday	Friday	Saturday
Sales (£)	535	418	502	289	612	876

a) Plot these points on a graph and join them with a straight line.

b) Is it sensible to read between Monday and Tuesday?

c) If the shop wanted to open on Sunday, could you tell them what their likely sales will be?

EXERCISE 9.2B

1 Julie carried out a survey of how many people were in each car passing the school gates between 08:30 and 09:00.

Here is the graph of her results.

Why is it not sensible to join the tops of the vertical lines?

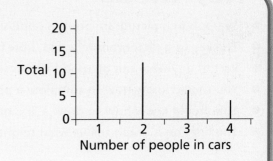

Exercise 9.2B cont'd

2 Eddie makes a cup of tea and measures the temperature as the tea cools down. Here are his measurements:

Time, seconds	0	20	40	60	80	100	120
Temperature °C	100	92	76	46	28	16	10

a) Plot these points on a graph and join them with straight lines.
Use your graph to find out:

b) what the temperature was after 50 seconds

c) how long it took the tea to cool down to 85°C.

3 The amount of chemical produced in an industrial process is shown below.

Time	08:00	10:00	12:00	14:00	16:00	18:00
Amount (kg)	0	24	52	91	112	124

a) Plot these points on a graph and join them with a straight line.

b) How much would be produced at 11:30 and 17:00?

c) Where is the greatest amount produced in a two-hour period?

d) By what time would 80 kg be produced?

Key ideas

- Symbols in a pictogram are the same size and are placed in rows and columns.
- The key to a pictogram tells you 'how many' each symbol represents.
- Part of a symbol can be used to represent a fraction of the number.
- You should know how to complete a pictogram.
- You should know how to draw a line graph and use it to answer questions.
- The lines on a graph tell us what might happen between the plotted points.

10 Listing

Three friends, Anne, Bill and Chloe, go to the cinema. The film is so popular that there are just three seats left, all next to each other. The friends can't decide who will sit where. Write down the different ways that they can sit next to each other.

Exam tip

Use letters for the friends. This will make listing easier.
Try to use some systematic way of listing. For example, keep A in seat 1 and change the others. Then keep B in seat 1, etc.

| A B C | B A C | C A B |
| A C B | B C A | C B A |

There are 6 possible ways they can sit.

EXERCISE 10.1A

1   How many 2-digit numbers can you make using the digits 5 and 7?

2   Dave, Emma and Fred sit in three seats next to each other in assembly. Emma insists on sitting at one end. List different ways they can sit.

3   I have three jumpers. Two are yellow and one is black. I wear a different jumper each day for three days. How many different orders are there in which to wear the jumpers?

4   List the different ways of arranging the letters, P, Q, R and S.
    (**Hint:** There are 24 different ways.)

5   I have a box of green and red counters. How many different patterns can I make using three counters?

EXERCISE 10.1B

1 I have a box of red counters and a box of blue counters. How many different two-counter patterns can I make?

2 How many different 3-digit numbers can you make using the digits 1, 2 and 3?

3 A school has the same three meals on offer each day: Pizza, Burger and Salad. List all possible selections of meal you can have in three days, for example: Pizza, Salad, Pizza.

4 Four people, A, B, C and D, work in an office. They must have their annual holiday one at a time. D wants to be last and A and B want to follow on after each other. List the possible orders of the holidays.

5 How many 4-counter patterns can you make with green and red counters only?

Key ideas

- When making lists it is easier to use letters or some other shorthand to represent the items.
- A systematic approach is best to ensure that all possible arrangements have been found.

Revision exercise

1 The pictogram below represents the income of an ice cream seller during one week in July. Copy and complete the pictogram.

‾‾‾ represents £10.

Day of the week	income (£)
Monday	
Tuesday	
Wednesday	60
Thursday	
Friday	45
Saturday	
Sunday	85

a) On which day did he sell most ice creams?

b) How much more money did he take on Wednesday than Tuesday?

c) How much money did he take altogether?

d) Give one possible reason why he sold so little ice cream on Thursday and one possible reason why he sold so much on Saturday.

2 This pictogram shows the results for a local football team for one season.

The symbol ◖ represents 2 matches

win ○ ○ ○ ○ ◔
draw ○ ○ ◖
lose ○ ◸

a) How many matches were played altogether?

b) Using the same symbols, draw a pictogram to show the following results for a team.

win: 8 matches; **draw:** 7 matches; **lose:** 3 matches

3 The table below shows the width of a tree as it grows.

Age (years)	10	20	30	40	50
Width (cm)	10	16	25	41	63

a) Plot these points on a grid and join them.

b) Use your graph to find out
  (i) the width of the tree after 35 years
  (ii) how old the tree was when its width was 20 cm and when its width was 50 cm.

4 Elaine has two skirts, one blue and one red. She has three blouses, one grey, one yellow and one white. How many different outfits can she wear?

5 Mum gives us a choice for Sunday lunch.

Meat	Beef, Pork, Chicken
Potatoes	Roast, Boiled
Vegetables	Peas, Cabbage

How many different meals can we choose? Only one item can be chosen from each list.

Stage 2

Chapter 1	**Using decimals**	**137**
Chapter 2	**Sequences**	**141**
	Patterns	141
	Rules	143
	Finding *n*	145
Chapter 3	**Drawing angles**	**151**
	Measuring angles	151
Chapter 4	**Multiplication and division**	**162**
	Multiplying larger numbers	162
	Dividing larger numbers	164
Revision exercise A1		**167**
Chapter 5	**Mass or weight**	**169**
Chapter 6	**Probability**	**172**
Chapter 7	**Percentages**	**178**
	Pie chart scales	181
Chapter 8	**Time**	**185**
Revision exercise B1		**190**

CONTENTS

Chapter 9	**Estimating lengths and angles**	**191**

| *Chapter 10* | **Solids** | **199** |
| | Making 3D shapes | 199 |

| *Chapter 11* | **Median and mode** | **204** |

| *Chapter 12* | **Formulae** | **208** |
| | Formulae in words | 208 |

| *Revision exercise C1* | | **213** |

Chapter 13	**Drawing reflections**	**215**
	Reflection writing	216
	Simple reflection	218
	More difficult reflections	221
	Planes of symmetry	223

| *Chapter 14* | **Maps and plans** | **227** |
| | Using maps | 231 |

| *Chapter 15* | **Two-way tables** | **236** |

| *Revision exercise D1* | | **241** |

1 Using decimals

You should already know

- what decimal numbers represent, e.g.

Hundreds	Tens	Units	.	Tenths	Hundredths	
	4	1	.	6	2	$= 41 + \frac{6}{10} + \frac{2}{100} = 41\frac{62}{100}$
3	2	1	.	1	2	$= 321 + \frac{1}{10} + \frac{2}{100} = 321\frac{12}{100}$
		1	.	0	6	$= 1 + \frac{6}{100} = 1\frac{6}{100}$

- how to change between units.
  The table shows the connection between some metric units. Use these to complete the Activity.

Length

1 kilometre (km)	=	1000 metres
1 metre (m)	=	100 centimetres or
		1000 millimetres (mm)
1 centimetre	=	10 millimetres

Mass

1 kilogram (kg)	=	1000 grams (g)
1 tonne (t)	=	1000 kilograms

Capacity

1 litre (l)	=	1000 millilitres (ml) or
		100 centilitres (cl)

ACTIVITY

Copy these tables and complete them by changing the figures in the shaded boxes into the other units.
The first line of each table has been done for you.

mm	cm	m	km
40 000	4000	40	0·04
			2·5
		1200	
6800			

g	kg	tonne
250 000	250	0·25
		1·2
	30 400	
100 000		

ml	cl	l
2000	200	2
		4·5
3000		

Whenever you add or subtract money you are working with decimals.
Remember to line up the decimal points.

EXAMPLE 1

a) £3·62 + £4·14   **b)** £4·93 + £5·48
c) £5·36 − £3·24   **d)** £4·25 − £2·85

a) 3·62
     + 4·14
     ‾‾‾‾‾
     7·76   Answer £7·76

b) 4·93
     + 5·48
     ‾‾‾‾‾
     10·41   Answer £10·41

c) 5·36
     − 3·24
     ‾‾‾‾‾
     2·12   Answer £2·12

d) 4·25
     − 2·85
     ‾‾‾‾‾
     1·40   Answer = £1·40

Exam tip

When working with money you should write the answer as £1·40 not £1·4.

EXAMPLE 1

Add 2 metres and 130 cm.

130 cm = 1·30 metres so 2 metres + 130 cm = 2 m + 1·3 m = 3·3 m

Exam tip

Units of measurements should always be the same.

EXERCISE 1.1A

1 Put these quantities in order, smallest first:
   a) £2, 325p, £3·15, 35p, £13·51, £1·25
   b) 4 m, 2000 cm, 0·15 km, 100 mm, 5 cm, 2·5 m
   c) 0·5 kg, 505 g, 1·5 kg, 5 kg, 0·05 tonne
   d) 20 cl, 2 l, 240 ml, 0·5 l, 4500 ml, 45 cl

2 Work out the answers to these:
   a) £1·40   **b)** £8·50   **c)** £7·35   **d)** £12·30   **e)** £7·54   **f)** £8·15
      + £3·35   + £4·36   + £4·70   − £4·20   − £2·63   − £5·20

3 Work out the answers to these:
   a) £2·67 + 184p   **b)** £24 − 868p   **c)** £143·30 − £9·99

4 Work out the answers to these:
   a) 8·4 m + 1·2 m   **b)** 2 m + 5 m 30 cm   **c)** 7 m 18 cm + 6 m 24 cm
   d) 3 m + 435 cm   **e)** 5 m 50 cm − 2 m   **f)** 4 m − 1·2 m
   g) 6000 cm − 4 m 3 cm

Exercise 1.1A cont'd

5 Work out the answers to these:

    a) 3 kg + 400 g   **b)** 5·2 l + 3·4 l   **c)** 600 g + 3333 g (answer in kg)

    d) 4200 g − 2 kg   **e)** 5·64 l − 1600 ml   **f)** 13·64 kg − 4·93 kg

6 The sign on a bridge states 'maximum height 3·42 m'.

    Which of these lorries could go through and what would be the distances between the top of each lorry and the bottom of the bridge?

    a) height of lorry = 3600 cm

    b) height of lorry = 2·4 m

    c) height of lorry = 3·2 m

7 Find the lengths marked?.

    a)

    b)
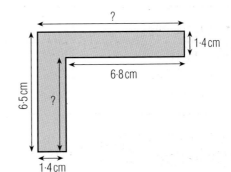

EXERCISE 1.1B

1 Put these quantities in order, smallest first:

    a) £5, 537p, £5·38, 53p

    b) 5 m, 5000 cm, 0·5 km, 500 mm

    c) 0·3 kg, 303 g, 1·3 kg, 0·3 tonne

    d) 30 cl, 3 l, 30 ml, 0·33 l

2 Work out the answers to these:

    a) £3·67   **b)** £9·56   **c)** £8·55   **d)** £14·90   **e)** £8·59   **f)** £7·75

      + £4·24   + £6·33   + £1·79   − £6·70   − £3·33   − £6·20

3 Work out the answers to these:

    a) 52p + £1·10   **b)** £1·29 − 40p   **c)** 32p + £1·78

Exercise 1.1B cont'd

4 Work out the answers to these:
    a) 9·3 m + 3·2 m   **b)** 5 m + 9 m 90 cm   **c)** 9 m 18 cm + 16 m 14 cm
    d) 6 m + 555 cm   **e)** 3 m 90 cm − 2 m   **f)** 3 m − 1·9 m
    g) 900 cm − 6 m 8 cm

5 Work out the answers to these:
    a) 8 kg + 900 g   **b)** 9·2 l + 1·9 l   **c)** 800 g + 4444 g (answer in kg)
    d) 5500 g − 2 kg   **e)** 6·54 l − 1700 ml   **f)** 23·69 kg − 5·95 kg

6 Find the lengths marked?
    a)

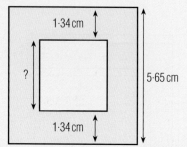

    b)

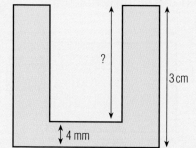

Key ideas

- To change from km into metres multiply by 1000.
- To change from litres into millilitres multiply by 1000.
- To change from kg into grams multiply by 1000.
- Write amounts of money with two decimal places, e.g. £4·20 not £4·2.

2 *Sequences*

Patterns

EXAMPLE 1

Look at this pattern:

What is the next shape?

A pattern like this is called a sequence.

The third term of this sequence is

EXAMPLE 2

Another sequence is:

Draw the next two terms in the sequence.

When you count the triangles in each pattern you get

1 4 9 and this is a sequence of numbers.

ACTIVITY 1

Write down the next three terms in the sequence in Example 2 above.

Do you recognise the numbers?

EXERCISE 2.1A

1 Write down the next three terms in each of these sequences.

   a) 2, 4, 6 –, –, –   **b)** 5, 10, 15, –, –, –   **c)** 11, 9, 7, –, –, –

2 Draw the next three shapes in this sequence of square tiles.

   Write down the first six terms of the sequence of the number of tiles.

3 This is a pattern of matchsticks arranged to look like rockets.

   a) Draw the fourth rocket.

   b) Write down the first six terms of the sequence of the number of matchsticks.

1 Write down the next three terms in each of these sequences:

a) 3, 7, 11, –, –, –　　**b)** 10, 13, 16, –, –, –　　**c)** 960, 970, 980, –, –, –

2 Here is a pattern of tiles.

a) Draw the next two shapes.

b) Write down the first six terms of the sequence of the number of squares.

3 Here is another pattern of tiles.

a) Draw the next two patterns.

b) Write down the first six terms of the sequence of the number of squares.

Rules

Look at this sequence:　　2, 4, 6, 8

The connection, or rule, between the terms is 'add 2'.

$$2 \xrightarrow{+2} 4 \xrightarrow{+2} 6 \xrightarrow{+2} 8$$

For this sequence:　　2, 4, 8, 16

the connection, or rule, is 'times 2'

$$2 \xrightarrow{\times 2} 4 \xrightarrow{\times 2} 8 \xrightarrow{\times 2} 16$$

Here is a collection of numbers.

Choose three of the numbers that you think form a simple sequence. Write them down and make a note of the rule.

Ask your neighbour to see if they can spot the rule.

Try it again with a different set of three numbers. (There are several different sequences for you to find.)

12　　3　　9

5　　　　　6

　　1

7　　　　18

EXERCISE 2.2A

1 Write down the next three terms in these sequences and the rule you used to find them.

    a) 1, 5, 9, 13, –, –, –   **b)** 19, 17, 15, 13, –, –, –

    c) 1, 3, 9, 27, –, –, –   **d)** 48, 24, 12, –, –, –

2 Write down the next three terms in these sequences and the rule you used to find them.

    a) 5, 10, 20, –, –, –   **b)** 1, 10, 100, –, –, –

    c) 8, 6, 4, 2, –, –, –   **d)** ⁻2, 0, 2, 4, –, –, –

EXERCISE 2.2B

1 Write down the next three terms in these sequences and the rule you used to find them.

    a) 7, 14, 21, 28, –, –, –   **b)** 21, 24, 27, 30, –, –, –

    c) 0·5, 1·0, 1·5, –, –, –   **d)** 0·2, 0·4, 0·6, –, –, –

2 Write down the next three terms in these sequences and the rule you used to find them.

    a) 128, 64, 32, –, –, –   **b)** 81, 27, 9, –, –, –

    c) 10, 21, 32, 43, –, –, – **d)** 8, 13, 18, 23, –, –, –

Finding n

Matchstick squares

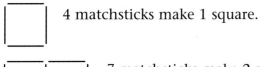 4 matchsticks make 1 square.

 7 matchsticks make 2 squares.

10 matchsticks make 3 squares.

How many matchsticks are needed for 4 squares?

Number of squares		Number of matchsticks
1	$\longrightarrow$	4
2	$\longrightarrow$	7
3	$\longrightarrow$	10

number of matchsticks = (number of squares × 3) + 1

$$n = (s \times 3) + 1$$

so for 4 squares $n = 4 \times 3 + 1 = 13$

We sometimes refer to term number or 'position' as well as the term value. In this example the term number is the number of squares and the term value is the number of matchsticks.

Term number, or position	Term
1	4
2	7
3	10

EXERCISE 2.3A

1 Look at this sequence: 8, 12, 16, ...

   a) What is the rule for getting the next term?

   b) What term or position number has a term value of 40?

2 Alan works on Saturdays washing cars for a garage.

Each Saturday he earns £20.

   a) Write down a sequence to show how much he has earned after one week,
      two weeks, etc.

   b) Work out a formula for the amount, £A, he has earned after N weeks.

3 The diagram shows a different arrangement of matchsticks making squares.

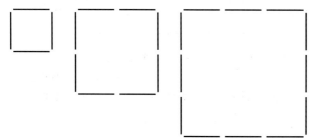

   a) How many matchsticks are used in the first square?

   b) Draw a table showing the number of matchsticks used for the first three squares.

   c) find a rule linking the number of matchsticks and the size of the squares.

Exercise 2.3 A cont'd

4   The diagram shows a fence made with posts and chains.

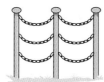

    a) How many chains will be needed for five posts?

    b) Fill in the missing numbers in this table:

Number of posts (*P*)	1	2	3	4	5	6	7
Number of chains (*C*)	0	3	6				

    c) Write down how the number of chains increases each time.

    d) Write down a rule which will help you find the **total** number of chains needed to make a fence when you already know the number of posts.

    Let *C* = total number of chains, and *P* = the number of posts.

EXERCISE 2.3B

1 Here is a pattern made with shaded tiles and white tiles.

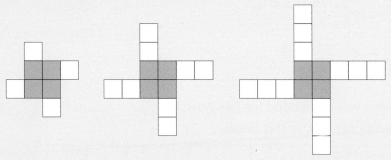

  a) How many shaded tiles are there in each shape?

  b) How many white tiles are there in each shape?

  c) Write down a rule for finding out the number of white tiles in each shape.

  d) How many tiles altogether will there be in the seventh shape?

2 Rose bushes are planted in a park so that every white rose (❀) is surrounded by four red roses (✿).

There are various ways of planting the bushes. Two of the ways are shown below:

Method A

Method B

  a) How many red roses are needed when six white roses are planted using method A?

  b) Twenty red rose bushes are planted using method B. How many white roses are needed?

  c) Which pattern, in general, uses fewer red roses? Explain your answer.

Something extra to try

You can sometimes find a rule by looking at what happens to the term number and the value of the term, e.g.

	Term or position number	Term
	1	4
goes up 1		goes up 3
	2	7
goes up 1		goes up 3
	3	10

The value of the term increases by 3 when the position number goes up by 1. This tells you that first you must multiply the position number by 3, and then add or subtract a fixed number to get the value of the term. So, looking at the first numbers:

$$1 \times 3 = 3.$$

We need to get an answer of 4 so we must add 1.

$$1 \times 3 + 1 = 3 + 1 = 4 \quad \text{which is correct.}$$

Check: $2 \times 3 + 1 = 6 + 1 = 7$ which is the correct answer for the second row

So the rule is term value = 3 × position number + 1.

Example:

Find the rule for this sequence 5, 8, 11, ...

	Term or position number	Term
	1	5
goes up 1		goes up 3
	2	8
goes up 1		goes up 3
	3	11

The rule is 3 × position number + 2, because 3 × 1 = 3 and we need to add 2 to get the answer 5.

Check: the third value is 11, and 3 × 3 + 2 = 11, so the rule works.

EXERCISE 2.4A

Use this method to find the rule for the following sequences:

    1 5, 9, 13, 17
    2 7, 12, 17, 22
    3 8, 14, 20, 26
    4 7, 6, 5, 4
    5 20, 30, 40, 50

EXERCISE 2.4B

Use this method to find the next two terms of these sequences. Explain how you found them.

    1 4, 8, 12, 16
    2 7, 9, 11, 13
    3 55, 45, 35, 25
    4 20, 27, 34, 41
    5 25, 23, 21, 19

Key idea

- Work out rules for sequences by looking at the differences between terms.

3 Drawing angles

You should already know

● how to read the scales on rulers.

Measuring angles

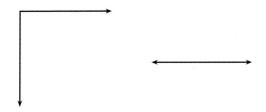

To measure the angle required to rotate one of the above arrows onto the other measurements, terms like '$\frac{1}{4}$ turn' and '$\frac{1}{2}$ turn' are accurate enough.

For angles such as these, however, a more accurate measurement is required.

To do this we use a scale marked in **degrees**.

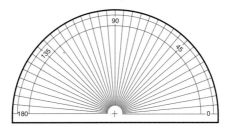

On this scale one whole turn is equal to 360 degrees. This is written as 360°.

So $\frac{1}{2}$ turn is equal to 180° and $\frac{1}{4}$ turn is equal to 90°.

EXAMPLE 1

What angle in degrees is equal to **a)** two whole turns

b) $\frac{3}{4}$ turn?

a) Two whole turns = $2 \times 360 = 720°$.

b) $\frac{3}{4}$ turn = $\frac{3}{4} \times 360° = 270°$.

Angles less than $\frac{1}{4}$ turn (90°) are called **acute** angles.

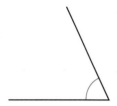  

Angles between $\frac{1}{4}$ (90°) and $\frac{1}{2}$ (180°) are called **obtuse** angles.

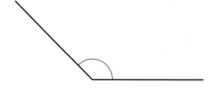 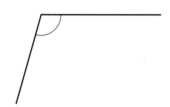

Angles of $\frac{1}{4}$ turn (90°) are called **right** angles.

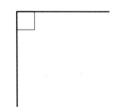 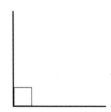

Angles of more than $\frac{1}{2}$ turn (180°) are called **reflex** angles.

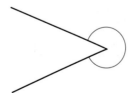 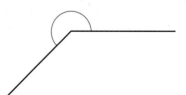

EXERCISE 3.1A

1 Are these angles acute, right, obtuse or reflex?

a)   **b)**   **c)**   **d)**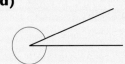

2 Are angles of these sizes acute, obtuse or reflex?
  a) 145°   **b)** 86°   **c)** 350°   **d)** 190°

EXERCISE 3.1B

1 What sort of angles are these?

a)   **b)**   **c)**

d)   **e)**

2 What sort of angles have these sizes?
  a) 126°   **b)** 226°   **c)** 26°   **d)** 90°

The instrument used to measure an angle is called a **protractor** or angle measurer. Most protractors are semi-circular in shape and can be used to measure angles up to 180°. Some protractors are full circles and can be use to measure angles up to 360°.

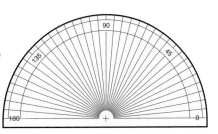

Chapter 3   *Drawing angles*

In this section you will find out how to use a semi-circular protractor to measure reflex (greater than 180°) angles. Since these are cheap, it is worth buying one for yourself.

EXAMPLE 2

Measure this angle in degrees.

Place your protractor so that the bottom line is along one of the arms of the angle and the centre is at the point of the angle.

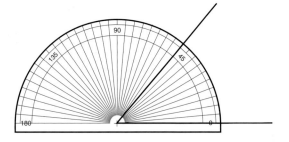

You will notice that there are two scales round the outside of your protractor. Make sure that you use the one that starts with zero. In this case it is almost certainly the inside one.

Go round this scale until you reach the other arm of the angle. You should be able to read off the scale at 50°.

So the angle is 50°.

Exam tip

One of the most common errors in measuring scales is to choose the wrong one of the two scales. Make sure you use the one that starts at 0. A useful further check is to estimate the angle first. If the angle is obviously **acute** (less than $\frac{1}{4}$ turn), then the angle is less than 90°. If the angle is obviously obtuse (greater than $\frac{1}{4}$ turn but less than $\frac{1}{2}$ turn) then the angle is between 90° and 180°. Knowing approximately what the angle is should prevent you using the wrong scale.

EXAMPLE 3

Measure this angle in degrees.

Place your protractor so that the bottom line is along one of the arms of the angle and the centre is at the point of the angle.

In this case use the scale where the zero is on the left. It is almost certainly the outside one.

Go round this scale until you reach the other arm of the angle. You should be able to read off the scale at 150°.

So the angle is 150°.

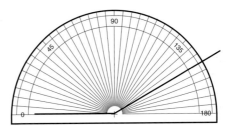

EXAMPLE 4

Measure this reflex angle in degrees.

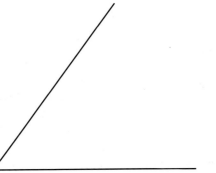

Example 4 cont'd

Measure the acute angle first.

This should read 53°.

Since the acute angle and the reflex angle together make one whole turn, they must add up to 360°.

So the reflex angle = 360° − 53° = 307°.

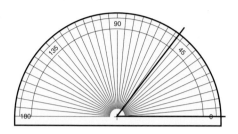

EXERCISE 3.2A

1 Work out the following angles in degrees.

    a) $1\frac{1}{2}$ turns   **b)** 3 turns   **c)** $\frac{1}{8}$ turn

2 Copy and complete this table by

    (i) estimating each angle   **(ii)** measuring each angle.

	Estimated angle	Measured angle
a)		
b)		
c)		
d)		
e)		
f)		
g)		
h)		
i)		
j)		

  a)

  b)

Exercise 3.2A cont'd

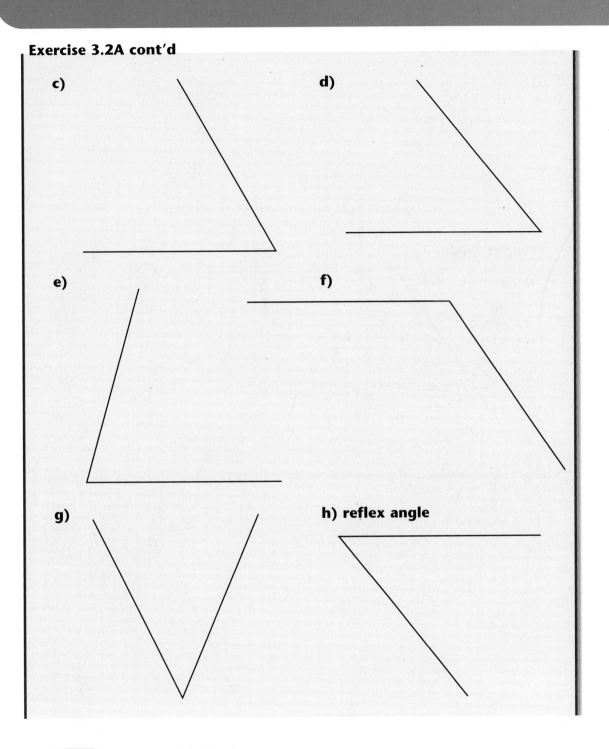

c)

d)

e)

f)

g)

h) reflex angle

Exercise 3.2A cont'd

i)

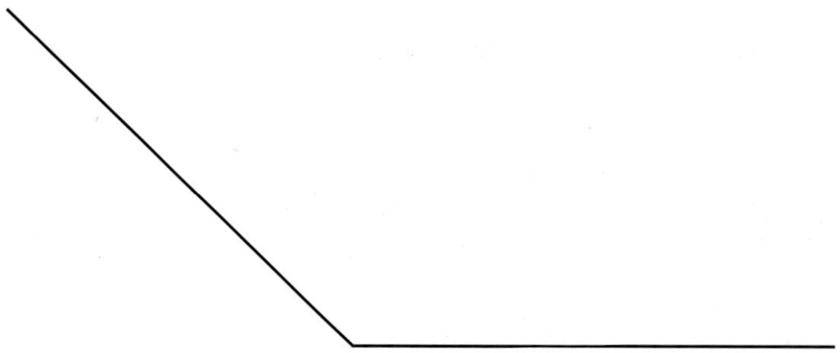

j) reflex angle

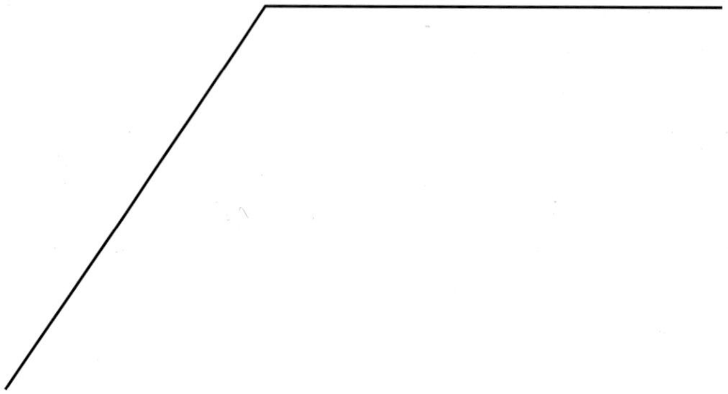

EXERCISE 3.2B

1   Work out the following angles in degrees.

   a) $\frac{1}{3}$ turns   **b)** $\frac{2}{3}$ turns   **c)** $\frac{1}{6}$ turn

2   Copy and complete this table by
    (i) estimating each angle   **(ii)** measuring each angle.

	Estimated angle	Measured angle
a)		
b)		
c)		
d)		
e)		
f)		
g)		
h)		
i)		
j)		

a)

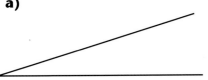

b)

c)

d)

Chapter 3 *Drawing angles*

Exercise 3.2B cont'd

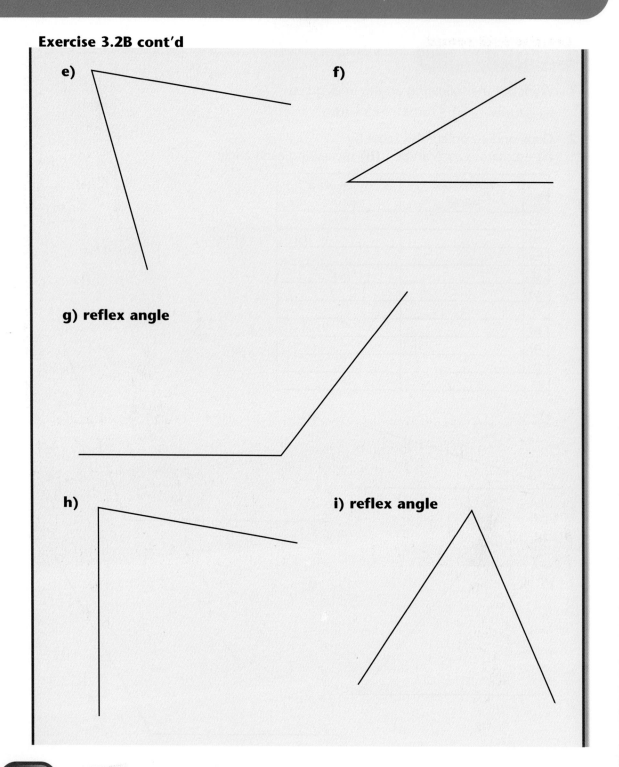

e)

f)

g) reflex angle

h)

i) reflex angle

Exercise 3.2B cont'd

j)

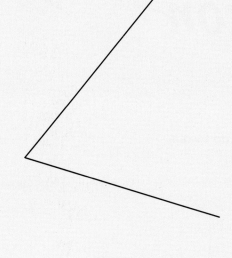

Key ideas

- Acute angles are less than 90°, obtuse angles are between 90° and 180° and reflex angles are greater than 180°.
- A full turn = 360°, a $\frac{1}{2}$ turn = 180° and a $\frac{1}{4}$ turn = 90°.

4 Multiplication and division

You should already know

- how to add and subtract numbers
- how to multiply and divide with numbers up to ten
- how to multiply and divide by 10.

Multiplying larger numbers

Look at this pattern.

$1 \times 4 = 4$
$2 \times 4 = 8$
$3 \times 4 = 12$
$4 \times 4 = 16$
$5 \times 4 = 20$
$6 \times 4 = 24$
$7 \times 4 = 28$

$8 \times 4 = 32$
$9 \times 4 = 36$
$10 \times 4 = 40$

It's the 4-times table. Can you continue it?

$11 \times 4 = \text{.....................}$
$12 \times 4 = \text{....................}$

The next two numbers are 44 and 48.

EXAMPLE 1

Work out 43×4.

You could continue the pattern but it will take a long time and you may make a mistake.

Try another way. $43 = 40 + 3$

so $43 \times 4 = 40 \times 4 + 3 \times 4$

but 40 is 10×4

To work out 43×4 is the same as $10 \times 4 \times 4 + 3 \times 4$

$= 10 \times 16 + 12$

$= 160 + 12$

$= 172$

EXAMPLE 2

Work out 27×8.

$$27 \times 8 = 20 \times 8 + 7 \times 8$$
$$= 10 \times 2 \times 8 + 7 \times 8$$
$$= 10 \times 16 + 56$$
$$= 160 + 56$$
$$= 216$$

EXERCISE 4.1A

1	23×3	**6**	37×5
2	18×6	**7**	49×2
3	41×2	**8**	17×7
4	32×4	**9**	40×9
5	13×8	**10**	83×3

EXERCISE 4.1B

1	32×3	**6**	61×6
2	57×2	**7**	45×3
3	19×5	**8**	15×9
4	24×7	**9**	74×4
5	42×4	**10**	99×8

EXAMPLE 3

Work out 36×7

	Tens	Units	
		36	
	$\times$	7	
		42	7×6
	2	10	$7 \times 3 \times 10$
	2	52	add to give 36×7

EXAMPLE 4

Work out 84×9

	84	
$\times$	9	
	36	9×4
7	20	$9 \times 8 \times 10$
7	56	add to give 84×9

EXERCISE 4.2A

1	38×3	**6**	47×4
2	19×9	**7**	92×7
3	64×2	**8**	31×9
4	81×6	**9**	17×8
5	57×5	**10**	84×5

EXERCISE 4.2B

1	74×4	**6**	83×3
2	29×7	**7**	91×9
3	13×9	**8**	46×2
4	71×8	**9**	18×6
5	48×5	**10**	75×5

Dividing larger numbers

Here is part of the 8-times table

$3 \times 8 = 24$
$4 \times 8 = 32$
$5 \times 8 = 40$
$6 \times 8 = 48$
$7 \times 8 = 56$

These could be written

$24 \div 8 = 3$
$32 \div 8 = 4$
$40 \div 8 = 5$
$48 \div 8 = 6$
$56 \div 8 = 7$

EXAMPLE 5

Work out $72 \div 8$.
as $9 \times 8 = 72$, $72 \div 8 = 9$.

EXAMPLE 6

Work out $72 \div 4$.

Your 4-times table probably stopped at $10 \times 4 = 40$ – unless you used the long method in Example 1!

There is a clue to solving this problem in $10 \times 4 = 40$.

$40 \div 4 = 10$

Subtract 40 from 72, leaving 32.

$32 \div 4 = 8$

giving $72 \div 4 = (40 \div 4) + (32 \div 4)$
$= 10 + 8$
$= 18$

EXAMPLE 7

Work out $72 \div 3$.
Try the same method.
$72 - 30 = 42$ (bigger than $3 \times 10 = 30$)
$42 - 30 = 12$
giving $72 \div 3 = (30 \div 3) + (30 \div 3) + (12 \div 3)$
$= 10 + 10 + 4$
$= 24$

EXERCISE 4.3A

1	$64 \div 8$	**6**	$81 \div 9$
2	$64 \div 4$	**7**	$76 \div 4$
3	$63 \div 7$	**8**	$75 \div 5$
4	$39 \div 3$	**9**	$81 \div 3$
5	$46 \div 2$	**10**	$90 \div 6$

EXERCISE 4.3B

1	$56 \div 7$	**6**	$99 \div 9$
2	$48 \div 3$	**7**	$96 \div 6$
3	$90 \div 5$	**8**	$84 \div 3$
4	$42 \div 2$	**9**	$91 \div 7$
5	$84 \div 4$	**10**	$56 \div 4$

More division

ACTIVITY 1

Work out

$72 \div 3$

$75 \div 3$

$78 \div 3$

$81 \div 3$

You should have answers 24, 25, 26, 27.

What about $73 \div 3$?

and $74 \div 3$?

73 is bigger than 72, so the answer should be bigger than 24.

But 73 is smaller than 75, so the answer should be smaller than 25.

Now try 76, 77, 79 and 80.

EXAMPLE 8

18 people are going on a journey in taxis. Each taxi can take 4 people. How many taxis will be needed?

To find how many, work out $18 \div 4$.

$16 \div 4 = 4$, $20 \div 4 = 5$, so the answer is between 4 and 5.

But if there are only 4 taxis, 2 people will be left behind. So 5 taxis are needed.

EXAMPLE 9

Three friends buy a bag of sweets.

There are 47 sweets in the bag. They share them out. How many will each have?

To find each share, work out $47 \div 3$.

$47 = 30 + 17$

$17 = 15 + 2$

so $47 \div 3 = (30 \div 3) + (15 \div 3)$ and 2 over

$= 10 + 5$ and 2 over

$= 15$ and 2 over.

The amount left over is called the **remainder**.

EXAMPLE 10

Find the remainders from these divisions.

a) 47 ÷ 4   **b)** 88 ÷ 6

a) 47 = 40 + 4 + 3

    47 ÷ 4 = (40 ÷ 4) + (4 ÷ 4) + 3

            = 10 + 1 with remainder 3

            = 11 remainder 3.

EXERCISE 4.4A

1   **a)** 16 ÷ 3   **b)** 57 ÷ 6
     c) 29 ÷ 2   **d)** 84 ÷ 9
     e) 77 ÷ 4

2   £100 is shared between 7 people.
How many pounds will each get?
How much is left over?

3   I have to take 90 books upstairs. I
can only carry 8 at a time. How
many times must I go upstairs?

4   There are 88 people at a meeting.
There are 9 chairs in each row.
How many rows are needed?
How many rows will be full?

EXERCISE 4.4B

1   **a)** 27 ÷ 4   **b)** 64 ÷ 7
     c) 51 ÷ 2   **d)** 49 ÷ 6
     e) 83 ÷ 5

2   A minibus can carry 9 people. 80
people are going on a trip. How
many minibuses will be needed?

3   Mary is packing 40 lamps in
boxes. Each box holds 6 lamps.
How many boxes can she fill?
How many lamps will be left?

4   Five friends share a prize of £66.
How much will each get?

Key ideas

- To multiply a 2-digit number, split it into tens and units and multiply separately.
- To divide a number larger than in the tables, subtract 10 times the dividing number as many times as you can.
- If, when you divide, there is a remainder, decide whether you need to go to the next number above.

Revision exercise

1 Work out the answers to these:
  a) £3 − £2·15
  b) £5 + £3·71
  c) £3·42 − £2·18
  d) 5·82 m + 6·4 m
  e) 7·14 m + 2300 m
  f) 9·4 m − 4 m 7 cm

2 Julie is building a display made of piles of soup cans.

The diagram shows the start of her display.

pile 1   pile 2   pile 3

a) How many cans make pile 3?

_____

b) Draw pile 4.

c) Copy and complete this table

Pile number	1	2	3	4	5
Number of cans	1	3			

d) Julie says, 'I need 22 cans to make pile 7.' Is she correct? Show your working.

3 How many degrees are there in
  a) 2 complete turns   **b)** $\frac{1}{2}$ turn?

4 Measure these angles:
  a)

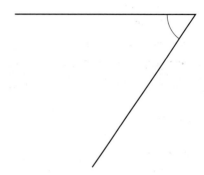

  b)

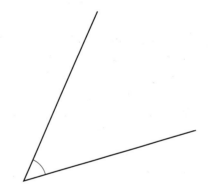

5 a)

b)

6 Work out
   a) 73×4 **b)** 54×8 **c)** 16×9

7 Work out
   a) $82 \div 2$ **b)** $84 \div 6$ **c)** $77 \div 8$

8 How many tins, each 6 cm wide, can fit in a single line on a shelf 100 cm long?

5 Mass or weight

Strictly speaking the proper name for how heavy something is is **mass,** but **weight** is the word commonly used.

Mass can be measured – either in **imperial** units – stones, pounds and ounces – or in **metric** units – grams and kilograms which are written as g and kg.

The imperial units for mass are rarely used nowadays. You will sometimes see ounces in recipes, pounds used for amounts of fruit and vegetables and stones for people's weights.

1 stone = 14 pounds and 1 pound = 16 ounces but you do not need to learn these facts.

The short version for pounds is lb.

Nearly all masses are now given in metric units and to have a good idea of a kilogram think of a bag of sugar which is 1 kilogram.

Small masses, like the sugar on a spoon, are weighed in grams and all larger weights are weighed in kilograms.

1 kilogram = 1000 grams (just like kilometres and metres), and you may need to change from one to the other.

169

EXAMPLE

a) Change the amounts in kilograms to grams.
   (i) 4 kg   **(ii)** 4·521 kg   **(iii)** 9·17 kg

b) Change the amounts in grams to kilograms
   (i) 7000 g   **(ii)** 4215 g   **(iii)** 82 034 g

a) (i) 4 × 1000 = 4000 g   1 kg = 1000 g, so multiply by 1000.
   (ii) 4·521 × 1000 = 4521 g   Same as **(i)**, move the decimal point three places to the right.
   (iii) 9·17 × 1000 = 9170 g   Same as above but add a zero so that the decimal point can be moved three places.

b) (i) 7000 ÷ 1000 = 7 kg   1 kg = 1000 g, so divide by 1000.
   (ii) 4215 ÷ 1000 = 4·215 kg   Same as **(i)**, move the decimal point three places to the left.
   (iii) 82 034 ÷ 1000 = 82·034 kg

EXERCISE 5.1A

1 In what metric units would you give the mass of
   a) yourself   **b)** a bicycle   **c)** a toffee   **d)** an orange?

2 Change these masses to grams.
   a) 9 kg   **b)** 1·129 kg   **c)** 3·1 kg   **d)** 0·3 kg   **e)** 0·012 kg.

3 Change these masses to kilograms.
   a) 2000 g   **b)** 1400 g   **c)** 3516 g   **d)** 94 652 g   **e)** 6600 g

4 Write these masses in order of size, lightest first.
   a) 4000 g   **b)** 52 000 g   **c)** 9·4615 kg   **d)** 874·12 g   **e)** 1·7 kg

5 John buys a bag of sugar weighing 1 kg, a bag of flour weighing 1·5 kg, a box of breakfast cereal weighing 450 gm and two tins of soup weighing 400 gm each.

   How much weight will he have to carry home?

6 Some old-fashioned types of kitchen scales used weights which were placed on balance pans as shown:

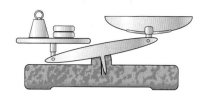

   The weights provided were of mass:

   1 g; 2 g; 2 g; 5 g; 10 g; 10 g; 20 g; 50 g; 100 g; 100 g; 200 g; 200 g; 500 g; 1 kg; 2 kg.

   Which of these weights would you use to weigh (i.e. balance) weights of:
   a) 127 g   **b)** 567 g   **c)** 1·283 kg   **d)** 2091 g   **e)** 2·807 kg.

EXERCISE 5.1B

1 In what metric units would you give the mass of
   a) £1 coin   **b)** a cow
   c) an exercise book   **d)** a packet of washing powder?

2 Change these masses to grams.
   a) 7 kg   **b)** 1·13 kg   **c)** 2·14 kg   **d)** 0·71 kg   **e)** 0·001 kg

3 Change these masses to kilograms.
   a) 8000 g   **b)** 6300 g   **c)** 5126 g   **d)** 49 612 g   **e)** 760 g

4 Write these masses in order of size, lightest first.
   a) 4123 g   **b)** 2104 g   **c)** 3·4165 kg   **d)** 0·174 kg   **e)** 2·79 kg

5 Farhat is making a chicken casserole.
   What is the total weight of her ingredients?
   Butter   50 g
   Bacon   125 g
   Onions $\frac{1}{4}$ kg
   Mushrooms $\frac{1}{4}$ kg
   Give your answer in kilograms.

6 What is $\frac{1}{2}$ kg in g?

Key idea

- There are 1000 g in a kg.

6 Probability

Probability uses numbers to show how likely an event is.

The probability of any event happening must lie between 0 and 1.

● **0 is the probability for an event which cannot happen.**
● **1 is the probability for an event which is certain to happen.**

You will already have used a probability line like this one:

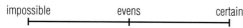

Adding numbers to this scale gives

but the line is usually shown with just the numbers

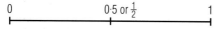

EXAMPLE 1

Here is a probability line

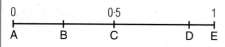

Match each of the events below with a letter on the probability line.

a) If you toss a coin it will come down tails.

b) If you roll a normal 6-sided die it will give a 7.

c) Christmas Day will fall on December 25th this year.

d) If you take a card from a pack of playing cards it will be a heart.

e) In England it will rain sometime during the next 14 days.

a) C   **b)** A   **c)** E   **d)** B   **e)** D

EXERCISE 6.1A

For each question make a copy of this probability line.

```
0              0·5            1
├──────────────┼──────────────┤
```

1 Show, with arrows, the probability

   a) of getting an odd number when you roll a normal 6-sided die

   b) that it will get dark tonight

   c) that you will swim the Channel.

2 Show, with arrows, the probability

   a) that it will rain every day in April

   b) that you will be late for school tomorrow

   c) that a Premier Club will win the FA Cup for the next five years.

3 Show, with arrows, the probability

   a) that it will be foggy tomorrow

   b) that it will snow in August

   c) of getting a 2 when you roll a 6-sided die.

4 If, without looking, you took one playing card from the five cards shown, what would be the probability of

   a) taking a red card

   b) taking a black card?

   Draw arrows on the probability line to show your answers.

EXERCISE 6.1B

For each question make a copy of this probability line.

```
0              0·5            1
├───────────────┼───────────────┤
```

1 Show, with arrows, the probability
   a) that you will win the National Lottery
   b) that the next dog you see will be black
   c) of getting a square number when you roll a roll a 6-sided die.

2 Show, with arrows, the probability
   a) that it will rain next Monday
   b) that it will snow in the Sahara Desert next August
   c) that the River Thames will freeze next week.

3 Show, with arrows, the probability
   a) that you will watch TV next week
   b) that there will be a general election within the next five years
   c) that your teacher will give you a bar of chocolate.

4 If you had a bag containing 5 red
   and 5 blue balls, draw arrows on
   your line to show the probability of

   a) picking a red ball without
   looking

   b) picking a yellow ball
   without looking.

Here is the probability line used in Example 1.

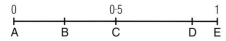

It is possible to give values to the positions on the line marked with letters.

Thus: A is at 0   C is at 0·5
         B is at 0·25 (halfway   D is at about 0·9
         between 0 and 0·5)   E is at 1

Note that it doesn't matter here whether D is actually at 0·8, 0·9, 0·95, etc. It is very likely that it will rain sometime during the next 14 days, but by no means certain, so the probability must be greater than 0·5 (i.e. evens), and less than 1 (i.e. certain). Experience suggests that the probability will be nearer to 1 than to 0·5!

What matters is that the number shows the approximate position on the line.

EXAMPLE 2

Copy this probability line and mark, with arrows, the points which show the probability of the following events happening. Next to each point write down the value of the probability.

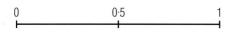

a) The Atlantic Ocean will freeze tomorrow.

b) The sun will shine on the first day of June.

c) You will get a number less than 3 when you roll a 6-sided die.

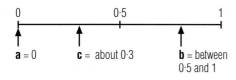

EXERCISE 6.2A

1 This probability line shows the probability of certain types of weather happening.

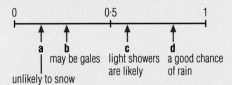

Write down the values of the probabilities for each weather type.

For questions 2 and 3 make a copy of this probability line.

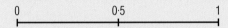

Draw arrows on your line to show your answers and write down the values for each probability.

2 Show the probability

   a) of getting a prime number when you roll a 6-sided die

   b) of getting a spade or a club card when you take one card from a pack of playing cards

   c) of getting a head or a tail when you spin an ordinary coin.

3 Show the probability

   a) of getting a number greater than 2 when you roll a 6-sided die

   b) that a new car will break down during the first year

   c) that you will experience an earthquake tomorrow.

EXERCISE 6.2B

For questions 1, 2 and 3, make a copy of this probability line.

```
0              0·5              1
├───────────────┼───────────────┤
```

Draw arrows on your line to show your answers and write down the values for each probability.

1 Show the probability

   a) of choosing a chocolate eclair from a bag of mint humbugs

   b) in a bag of 10 red counters and 2 blue counters taking out a red counter without looking

   c) you will have a drink tomorrow.

2 Show the probability

   a) that the first car you see on the road outside school will be red

   b) of rolling an ordinary 6-sided die and getting a number greater than 10

   c) that you will see an elephant tomorrow.

3 Show the probability

   a) of tossing a coin which lands and remains on its edge

   b) of flying to the moon before you are 20

   c) of taking an ace from a pack of playing cards.

Key ideas

- Probability uses numbers to show how likely an event is.
- The probability of any event happening must lie between 0 and 1.
  - 0 is the probability for an event which cannot happen.
  - 1 is the probability for an event which is certain to happen.
- Decimal numbers are used to show how likely or unlikely the occurrence of an event will be.

7 *Percentages*

You should already know

- how to multiply by $\frac{1}{4}$, $\frac{1}{2}$, $\frac{3}{4}$
- how to multiply using a calculator.

EXAMPLE 1

50% of cat owners say their cats prefer Frolic cat food.

50% means that 50 out of every 100 cat owners say their cats prefer **Frolic**.
You could show this in a 10 × 10 square like this:

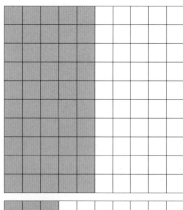

a) What fraction is shaded?

b) Write this fraction in its simplest form.

You can see that 50% is the same as $\frac{1}{2}$ so that half of the cat owners say their cats prefer **Frolic**.

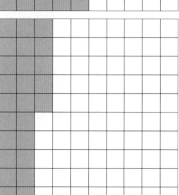

25% of dog owners feed their dogs with dry food.

Again, you can show this on a 10 x 10 square.

a) What fraction of dog owners use dry food?

b) Write the fraction in its simplest form.
You can see that 25% is the same as $\frac{1}{4}$.

c) What fraction of dog owners do not use dry food?

ACTIVITY

Write down $\frac{1}{2}$ of 44, 20, 162.

Write down $\frac{1}{4}$ of 20, 36, 400.

Write down $\frac{3}{4}$ of 20, 400, 64.

Write down 50% of 22, 12, 50.

Write down 25% of 200, 84, 120.

Write down 75% of 200, 12, 64.

EXERCISE 7.1A

1 Copy and complete this table.

Fraction	Decimal	Percentage
		25%
	0·5	
$\frac{3}{4}$		

2 Find 50% of the following amounts.

    a) £10   **b)** 120 cm   **c)** 20 kg   **d)** £1500

3 Find 75% of the following amounts.

    a) £36   **b)** 48 kg   **c)** 500 cm   **d)** £2000

    (**Remember**: 75% is the same as $\frac{3}{4}$ so find $\frac{1}{4}$ and then × by 3.)

4 'Books R Us' sell books at half the normal price, by post.

    You choose one or more books each month, but pay 25% of the cost for the post and packing.

    a) Jodi buys some books that cost £20. How much will the post and packing be?

    b) How much will it cost her altogether?

    c) Rachel buys three books which cost £8 each. How much will she have to pay altogether?

5 A test was marked out of 60. John gained 50%, Freda gained 25%.

    How many marks did each get?

6 Building regulations state that the width of a retaining wall should be no more than 50% of its height. What is the maximum width of a wall if the height is 5 m?

EXERCISE 7.1B

1 Work out 25% of these amounts.
   a) £12   **b)** £20   **c)** £4   **d)** £8·40

2 Find 75% of the following amounts.
   a) 48 kg   **b)** £16   **c)** £60   **d)** 24 m

3 Eddie used to get £4·50 each week for cleaning cars. He was given a 50% pay rise.
   a) How much extra did he get?
   b) What was his new wage?

4 In a clearance sale 'STARS ELECTRICAL DISCOUNT STORES' offer the following items with a 25% reduction.
   Copy and complete the table.

Item	Normal Price	Reduction	Sale Price
Washing machine	£400		
TV	£150		
Fridge/freezer	£240		
Microwave	£100		

5 A packet of rice states that it contains 25 % extra, and gives the usual mass as 300 g. How much extra rice is in the packet?

Pie chart scales

You need to be able to use a pie chart scale like the one shown here.

You can use the scales to find percentages of quantities shown in pie charts.

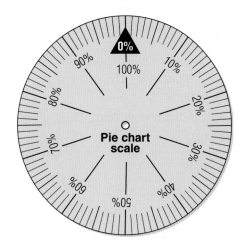

EXAMPLE 2

71% of the Earth's surface is water and only 29% is land.

This can be shown in a pie chart.

If you put the centre of the scale over the centre of the pie chart and line up 0% on the scale with the start of the water part (or sector) of the pie chart, then you can read off the percentages which are water and land. Check that you agree.

The water is found in the oceans, lakes and seas. Use your scale to read off the amount of water shown in each of the oceans in the pie chart below.

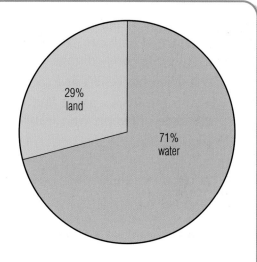

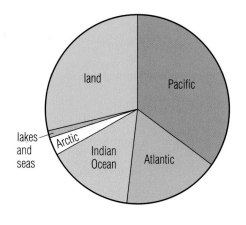

You may find it easier to rotate the scale about the centre so that the 0% line always lines up with the start of the sector you are working with.

EXERCISE 7.2A

1 This pie chart shows how Kevin spends a typical day. In parts **a)** and **b)** choose the answer you think is correct.

a) The percentage of time watching TV is
   (i) more than 25%
   (ii) 25%
   (iii) less than 25%.

b) The percentage of time sleeping is
   (i) more than 50%
   (ii) less than 50%
   (iii) 50%.

c) Use your pie chart scale to measure the actual percentages.

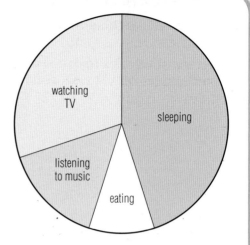

2 Joanne asked people which TV channel they watched between 10 p.m. and midnight one Friday night.

She showed her results in a pie chart.

Use your pie chart scale to find the percentage of people who watched each channel.

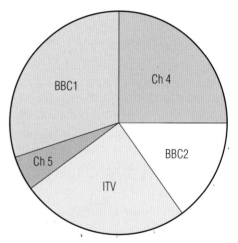

Exam tip

Always measure the percentages from zero – moving the scale round if necessary.

EXERCISE 7.2B

1 This pie chart shows how Perry spends a typical day.

Use your pie chart scale to work out what percentage of time Perry spent on each activity.

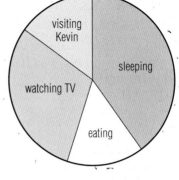

2 A school collected data on where its students went when they left Year 11. Here are the results.

Use your pie chart scale to work out what percentage of students followed each course.

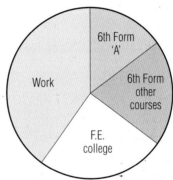

3 This pie chart shows the reading habits of a group of people.

Use your scale to work out what percentage of people read each newspaper.

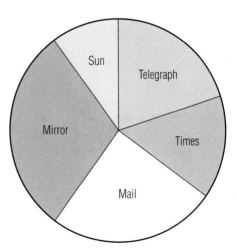

Exercise 7.2B cont'd

4   A class of 47 pupils were asked what sort of pet they had. The pie chart shows the results of the survey.

Use your pie chart scale to work out what percentage of pupils had each kind of pet.

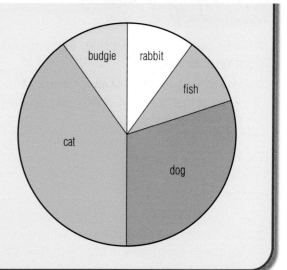

Key idea

● To work out sample percentages, use the fraction which is the same, e.g. for 25% use $\frac{1}{4}$.

8 Time

The time of the day can be given as a.m., p.m. or on the 24-hour clock. Most videos use the 24-hour clock and so you are most likely familiar with this.

However, you also need to be able to change from one to the other.

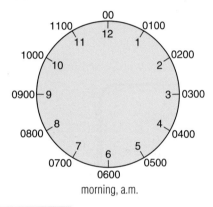

morning, a.m.

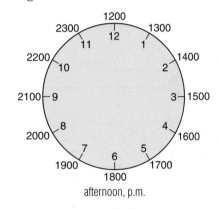

afternoon, p.m.

EXAMPLE 1

What are the following times on the 24-hour clock?

a) 9:14 a.m.   **b)** 11:32 a.m.

c) 1:30 p.m.   **d)** 6:55 p.m.

a) 0914   Times in a.m. are the same as on the 24-hour clock, except they must have four figures. In this case a zero is needed at the start.

b) 1132

c) 1330   To change p.m. to the 24-hour clock, add 12 to the hours and write without the colon (:).

d) 1855   Times in the 24-hour clock are often written with a colon (:). e.g. 18:55 in this case. Either way is acceptable.

EXAMPLE 2

What are the following times as a.m. or p.m.?

a) 0421   **b)** 1146   **c)** 1422   **d)** 2305

a) 4:21 a.m.   As it is in the morning just write as a.m., dropping the first zero.

b) 11:46 a.m.

c) 2:22 p.m.   As it is in the afternoon, subtract 12 hours.

d) 11:05 p.m.

The awkward times are those between midnight and 1 a.m. which start with 00 on the 24-hour clock, and between 12 noon and 1 p.m. from which 12 is not subtracted for p.m. times.

EXAMPLE 3

a) Write these times for the 24-hour clock.
   (i) 12:18 a.m.   (ii) 12:15 p.m.
b) Write these times as a.m. or p.m.
   (i) 0005   (ii) 1247.

a) (i) 0018   as it is between midnight and 1 a.m.
   (ii) 1215   as it is between midday and 1 p.m.
b) (i) 12:05 a.m. as it is between midnight and 1 a.m.
   (ii) 12:47 p.m. as it is between midday and 1 p.m.

When adding times, care has to be taken as there are 60 seconds in a minute and 60 minutes in an hour.

EXAMPLE 4

Mavis left for school at 7:45 a.m. and arrived 40 minutes later. At what time did she arrive?

7:45 + 40 minutes = 7 hours + 85 minutes   85 minutes is 1 hour and 25 minutes
= 8 hours 25 minutes = 8:25 a.m.

EXAMPLE 5

Carla set her video to start recording at 1520 and to stop at 1710. For how long was it recording?

1 hour 50 minutes.   The easiest way to answer this question is to count to the next hour, 1600. This is 40 minutes. Then, to get to 1710, a further 1 hour and 10 minutes are needed. So the total time is 1 hour 10 minutes + 40 minutes = 1 hour 50 minutes.

Exam tip

Most errors are made in time questions by using a calculator. It is best not to use a calculator. You must add the minutes separately and then change to hours and minutes.

An alternative way of finding the difference in times is to set the calculation out like this:

	Time difference in hours	Time difference in minutes
from 1520 until 1600		40
from 1600 until 1700	1	
from 1700 until 1710		10
Total	1	50

EXERCISE 8.1A

1 Write these times on the 24-hour clock.
   a) 3:45 a.m.   **b)** 7:53 a.m.   **c)** 11:44 a.m.   **d)** 6:42 a.m.   **e)** 12:04 a.m.
2 Write these times on the 24-hour clock.
   a) 3:25 p.m.   **b)** 9:50 p.m.   **c)** 11:04 p.m.   **d)** 7:30 p.m.   **e)** 12:33 p.m.
3 Write these times using a.m. notation.
   a) 1145   **b)** 0553   **c)** 0140   **d)** 0920   **e)** 0010
4 Write these times using p.m. notation.
   a) 1345   **b)** 1553   **c)** 2140   **d)** 2259   **e)** 1210
5 Write these times on the 24-hour clock.
   a) 1:50 a.m.   **b)** 2:40 p.m.   **c)** 11:49 a.m.   **d)** 6:30 p.m.   **e)** 12:02 a.m.
6 Write these times using either a.m. or p.m. notation.
   a) 0345   **b)** 1456   **c)** 2340   **d)** 1159   **e)** 1255
7 A train due at 1440 was 50 minutes late. What time did it arrive?

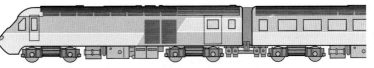

8 Peter went for a ride on his bike. He left home at 8:35 a.m. and was out for 1 hour 45 minutes.
   What time did he arrive back home?

Exercise 8.1A cont'd

9 Kylie set her oven to switch on at 3:45 p.m. and switch off at 6:20 p.m. How long was the oven on?

10 A coach left Barnsley at 1547 and arrived in Manchester at 1820. How long did it take?

EXERCISE 8.1B

1 Write these times on the 24-hour clock.

    a) 2:15 a.m.   **b)** 8:43 a.m.   **c)** 10:34 a.m.   **d)** 5:21 a.m.   **e)** 12:45 a.m.

2 Write these times on the 24-hour clock.

    a) 4:50 p.m   **b)** 7:20 p.m.   **c)** 10:49 p.m.   **d)** 8:44 p.m.   **e)** 12:01 p.m.

3 Write these times using a.m. notation.

    a) 1041   **b)** 0232   **c)** 0430   **d)** 1120   **e)** 0048.

4 Write these times using p.m. notation.

    a) 1440   **b)** 1723   **c)** 1940   **d)** 2019   **e)** 1203.

5 Write these times on the 24-hour clock.

    a) 3:20 a.m.   **b)** 2:08 p.m.   **c)** 12:49 a.m.   **d)** 9:35 a.m.   **e)** 11:02 p.m.

6 Write these times using either a.m. or p.m. notation.

    a) 0435   **b)** 1516   **c)** 2140   **d)** 0159   **e)** 1452

7 A ferry left Liverpool at 1420 and arrived at Belfast at 2105. How long did it take?

8 Pam left Lincoln at 7:50 a.m. and took 2 hours 45 minutes to reach Cambridge. What time did she arrive in Cambridge?

9 Joseph set his video recorder to switch on at 8:35 p.m. and switch off at 10:20 p.m. How long was the recorder on?

10 Chika left school at 1640, went to town shopping and arrived home $2\frac{1}{2}$ hours later. What time did she get home?

Exercise 8.1B cont'd

11 Part of the Howerdale railway timetable is given by:

Stanford 8:15 a.m.

Luker 9:05 a.m.

Melpett 9:40 a.m.

Haker 11:20 a.m.

Golpath 11:35 a.m.

Oldway 1:20 p.m.

a) How long does the train take to get from
  (i) Stanford to Melpett
  (ii) Haker to Oldway
  (iii) Luker to Golpath
  (iv) Melpett to Oldway?

b) If the train is delayed in Luker for 35 minutes, at what time will it arrive in
  (i) Golpath
  (ii) Oldway?

Key idea

● To add times, add the minutes separately and then change the total to hours and minutes. Finally, add in the hours.

B1 Revision exercise

1 Write

    a) 25·23 kg in g

    b) 14 080 gm in kg

2 Find the total weight of 1·43 kg, 500 g, 955 g, 23·54 kg

For questions 3 and 4 you will need to copy this probability line.

```
impossible          evens          certain
    ├───────────────┼───────────────┤
    0              0·5              1
```

3 On a copy of the line show the probabilities of the following events occurring.

    a) Valentine's Day will be on February 14th.

    b) You will watch TV tomorrow.

    c) It will rain during the next 7 days.

4 On a copy of the line show the probabilities of the following events occurring.

    a) The person next to you is female.

    b) You throw a die and get a 1.

    c) The number on the National Lottery bonus ball is larger than 6.

5 Brian left Derby at 6:25 a.m. and arrived in Cambridge 1 hour and 55 minutes later. What time did he arrive in Cambridge?

He left Cambridge at 1730 and arrived home at 1915. How long did he spend travelling back from Cambridge?

6 The pie chart below shows the colours of cars in a car park.

Use your pie chart scale to work out what percentage of cars were each colour.

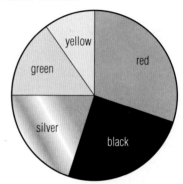

7 In a sale everything is reduced by 50%. How much will a suit originally priced at £110 sell for in the sale?

8 Put these in order, smallest first.

3 kg, 300 g, 1·4 kg, 2 g, 875 g, 2500 g

9 Write these times using the 24-hour clock.

6:20 a.m., 3:35 p.m., 11:23 p.m.

10 Write these times using the 12-hour clock.

1420, 1052, 2103

11 Find

    a) 50% of £54   **b)** 75% of 36 g

12 Find $\frac{1}{4}$ of 72.

13 A courier delivers a package worth £200. He charges a 25% handling charge. How much was the charge?

9 Estimating lengths and angles

You should already know

- how to identify types of angle

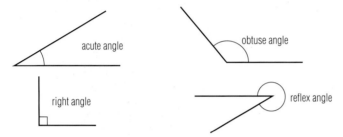

- how to use a protractor or angle measurer
- how to read the scales on a ruler.

EXERCISE 9.1A

1 This angle measures 30°.

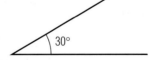

Estimate the size of the angles below. Write down your estimates and then check them using a protractor/angle measurer.

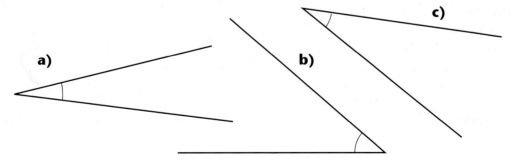

Exercise 9.1A cont'd

2 These angles measure 45° and 135° respectively.

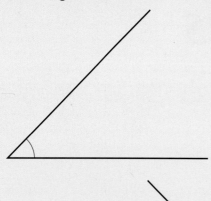

Estimate the size of the angles below. Write down your estimates and then check using a protractor/angle measurer.

a)

b)

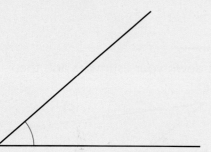
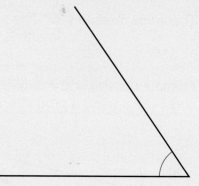

Exercise 9.1A cont'd

c)

d)

3 Estimate the size of the angles below. Write down your estimate and then check them using a protractor/angle measurer.

a)

Chapter 9 *Estimating lengths and angles*

Exercise 9.1A cont'd

b)

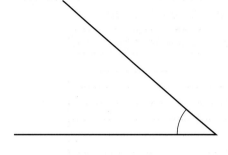

c)

d)

e)

Exercise 9.1A cont'd

Copy and complete the table for each of the angles, write down the type and estimated size of the angle and then measure the size.

Angle	Type	Estimated	Measured
a			
b			
c			
d			
e			
f			

4 This line is 10 cm long

10 cm

Estimate, and then check by measuring the lengths of the following lines

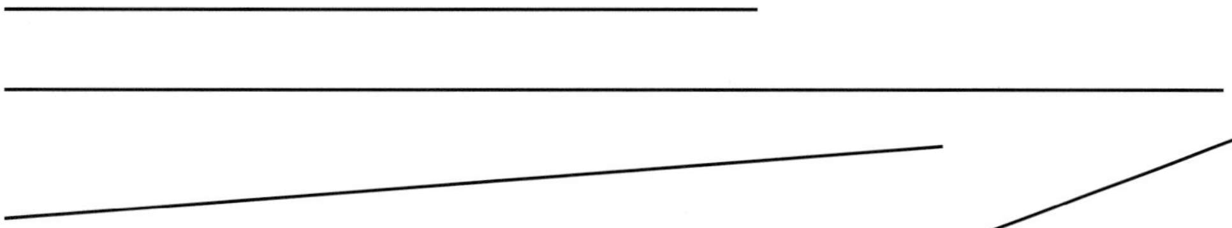

5 This line is 15 cm long

15 cm

Estimate the lengths of the sides of this rectangle.

195

Exercise 9.1A cont'd

6 Here is a sketch of a man, a house and a tree. If the man is about 2 m tall, estimate the height of the house and the tree.

EXERCISE 9.1B

1 This angle measures 60°.

Estimate the size of the angles below. Write down your estimates and then check using a protractor/angle measurer.

a)

b)

Exercise 9.1B cont'd

c)

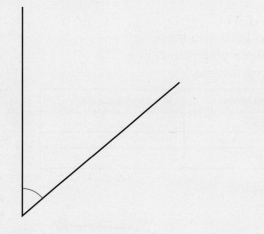

2 This angle measures 20°.

Estimate the size of the angles below. Write down your estimates and then check using a protractor/angle measurer.

a)

b)

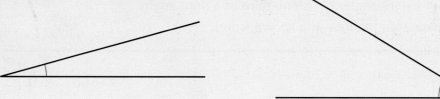

Exercise 9.1B cont'd

3 You need a 30 cm ruler.

Estimate the lengths of several objects and then check them by measurement. Show your results in a table. Three suggestions are given for you.

Object	Estimated	Measured
book		
width of desk		
pen or pencil		

4 This is a sketch map of a railway line.

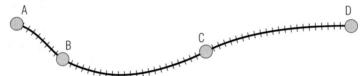

It is 20 km from A to B. Estimate the distance from B to C and A to D.

Key ideas

● 45° is half a right angle; 30° is one third of a right angle.

● To estimate a length, look at a known length first.

 **Solids**

You should already know

● the geometric names of common shapes, such as:
cube, cuboid, pyramid, cone, sphere, cylinder.

Making 3D shapes

A flat shape which can be folded to make
a 3D shape is called its **net**.

> ### Exam tip
> Give yourself practice in selecting and
> describing shapes to help you
> remember the words needed.

Here are two possible nets to make a cube.

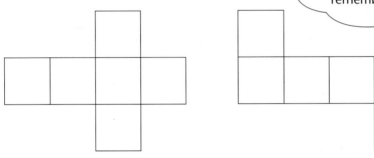

When actually making 3D shapes, flaps need to be added to edges so that they
can be glued together. These flaps are called tabs.

Tips on making 3D shapes:
● Put tabs on every other edge.
● Use card rather than paper, if possible, to construct a shape that will last!
● When using card, score the edges before folding.
● Draw the net as accurately as possible to obtain a good shape that fits
together well.
● Use glue that is suitable for the material you are using. If possible, use
quick-drying glue.
● Use the resulting shapes if you can – e.g. when suitably decorated, they can make
good gift/storage boxes.

Which of these is a possible net for a cube? Write Yes or No for each one.

a)

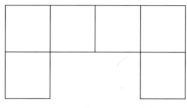

No

b)

Yes

EXERCISE 10.1A

1 Name each of these 3D shapes.

a)

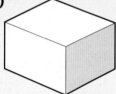

b)

c)

2 Name each of these 3D shapes.

a)

b)

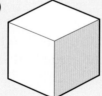

c)

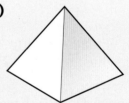

3 Sketch a net for a cube.

4 Draw a full-size net for this cuboid.

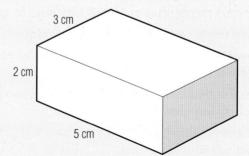

3 cm

2 cm

5 cm

Exercise 10.1A cont'd

5 When this net is folded to make a 3D shape, which vertex will join to N? Which line will join to CD? Which line will join to IH?

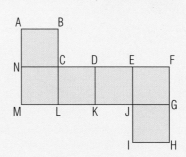

6 When this net is folded to make a 3D shape, which vertex will join to A? Which line will join to EF?

This solid is called a prism.

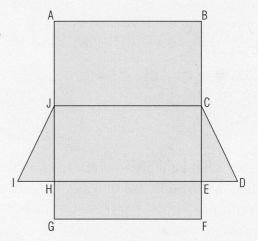

7 Sketch a copy of the net in question 5. Add tabs to some edges to make a net suitable for a model which needs gluing.

8 Sketch a copy of the net in question 6. Add tabs to some edges to make a net suitable for a model which needs gluing.

9 Copy this net full-size onto card and make the resulting solid.

This solid is called a tetrahedron.

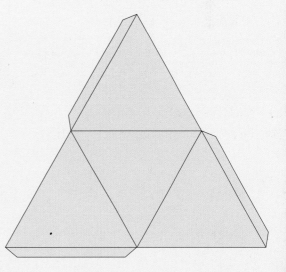

EXERCISE 10.1B

1 Name each of these 3D shapes.

a)

b)

c)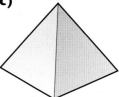

2 Name each of these 3D shapes.

a)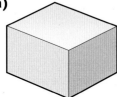

b)

c)

3 Sketch a net for a prism which has a triangular cross-section.

4 When this net is folded to make a 3D shape, which vertex will join to E? Which line will join to AN? Which line will join to LM?

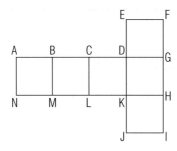

Exercise 10.1B cont'd

5 When this net is folded to make a 3D shape,
   (a) which vertex will join to S
   (b) which line will join to DE?

6 Sketch a copy of the net in question 5. Add tabs to some edges to make a net suitable for a model which needs gluing.

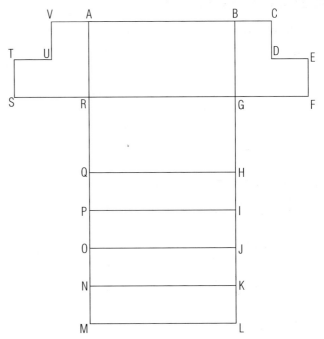

Key idea

● Common 3D shapes are cube, cuboid, prism, sphere, cylinder, pyramid, cone.

11 Median and mode

You should already know

- how to order numbers.

The median and the mode are two different types of 'average'.

EXAMPLE 1

Tariq and Farhat go ten-pin bowling.

These are their scores.

Tariq	7	8	5	4	7
Farhat	10	10	2	1	6

The **mode** is the most common score.

Tariq's mode is 7.

Farhat's mode is 10.

The **median** is the middle score when the scores are put in order.

Tariq	4	5	7	7	8
Farhat	1	2	6	10	10

The shaded scores are the middle scores.

Tariq's median score is 7; Farhat's median score is 6.

EXAMPLE 2

Here is a list of the weights of people in a 'Keep Fit' class.

73 kg, 58 kg, 61 kg, 43 kg, 81 kg, 53 kg, 73 kg, 70 kg, 73kg, 62 kg, 85 kg

Their instructor wants to know the average weight for the group.

One way is to put the weights in order and find the middle, that is the median, weight.

43 kg, 53 kg, 58 kg, 61 kg, 62 kg, 70 kg, 73 kg, 73 kg, 73 kg, 81 kg, 85 kg

These are two middle weights, 62 kg and 70 kg. The median is halfway between them, 66 kg.

The mode is 73 kg.

EXERCISE 11.1A

1 Find the median marks for each of these tests.

Remember to write the marks in order, smallest to largest, first.

a) 3   5   6   7   9

b) 8   9   4   3   7   3   1   7

c) 7   8   8   8   8   7   7   8   8   8   7

2 The time taken for a bus journey depends on the time of day.

Here are the times:

| 15 minutes | 7 minutes | 9 minutes | 12 minutes | 9 minutes |
| 19 minutes | 6 minutes | 11 minutes | 9 minutes | 14 minutes |

a) What is the median time for the journey?

b) What is the mode?

3 Twelve people have their handspan measured. The results are shown below.

225 mm	216 mm	188 mm	212 mm	205 mm
198 mm	194 mm	180 mm	194 mm	198 mm
200 mm	194 mm			

a) How many of the group had a handspan greater than 200 mm?

b) What are the median and mode?

4 The number of matches in ten different matchboxes was:

48   47   47   50   46   50   49   49   47   50

a) Find the median.

b) Find the mode.

Exercise 11.1A cont'd

5 The ages of a group of people are:

19, 23, 53, 19, 19, 16, 26, 77, 19, 27

Find the median and the mode.

EXERCISE 11.1B

1 Copy and fill in the table below for each of the sets of data:

Data set A	1, 1, 2, 3, 3, 3, 4, 4, 5, 6, 7
Data set B	1, 1, 2, 2, 3, 3, 3, 4, 5, 6, 7
Data set C	2, 2, 4, 4, 6, 6, 6, 8, 10, 12, 14

	Data set A	Data set B	Data set C
Median			
Mode			

b) Write down anything that you notice.

2 a) Find the median and mode of:

(i)	1	2	3	3	4	5
(ii)	10	20	20	30	70	
(iii)	110	120	120	130	170	
(iv)	7	10	13	16	19	

b) What do you notice about your answers to **(ii)** and **(iii)**?

3 In a survey a group of boys and girls wrote down how many hours of TV they watch each week.

Boys	17	22	21	23	16	12	15
	0	5	13	15	13	14	20
Girls	9	13	15	17	10	12	11
	9	8	12	14	15		

a) Find the median, mode and range for these figures.

b) Do the boys watch more TV than the girls?

Exercise 11.1B cont'd

4 The pay of ten workers in a small company is:

£10 000	£10 000	£10 000	£10 000
£13 000	£13 000	£15 000	£21 000
£23 000	£70 000		

a) Find the median and mode for the data.

b) Which of these averages do you think gives the best impression of the average pay?

5 The marks scored in a test were:

20	16	18	17	16	18	14	13
18	18	15	18	19	9	12	13

a) Find the median.

b) Find the mode.

6 A gardener measures the heights of a group of plants.

The heights were:

50 cm   65 cm   80 cm   40 cm   35 cm.

Find the median and the mode.

7 Seven people go to an evening class to learn how to paint with oils.

Their ages are:

18   19   17   45   37   69   23

Calculate the median and the mode.

Key ideas

● The median is the middle number of set of numbers arranged in order.

● The mode is the most common number.

12 *Formulae*

You should already know

- how to add, subtract, multiply and divide whole numbers and decimals
- how to add and subtract negative numbers.

Formulae in words

ACTIVITY

Here are some word formulae for you to use.

Length in millimetres is the length in centimetres multiplied by 10.

Distance travelled is equal to the speed multiplied by the time taken.

The amount in kilograms is roughly the amount in pounds divided by 2.

Use these formulae to change or calculate

a) 30 cm to millimetres

b) 10 pounds to kilograms

c) the distance travelled by a car at 40 miles per hour for 4 hours

d) 60 mm to centimetres

e) 40 pounds to kilograms

f) the time taken to travel 18 miles at 3 miles per hour.

EXAMPLE 1

To find the perimeter of a rectangle add the length and width and then multiply the total by 2.

Work out the perimeter of these rectangles.

a) length = 4 cm, width = 3 cm

b) length = 58 cm, width = 32 cm.

a) length + width = 4 + 3 = 7.
perimeter = 7 × 2 = 14 cm.

b) length + width = 58 + 32 = 90.
perimeter = 90 × 2 = 180 cm.

EXAMPLE 2

To work out the charge to hire a coach in £, Mrs Martin multiplies the number of miles by 2 and adds on 50.

How much does Mrs Martin pay to hire the coach to travel

a) 100 miles

b) 225 miles?

EXERCISE 12.1A

1 This is a rough rule for changing inches into centimetres.

> **To change inches into centimetres multiply by $2\frac{1}{2}$.**

    a) About how many centimetres is 2 inches?

    b) A foot is 12 inches. How many centimetres is this?

2 The time in minutes needed to cook a piece of beef is ...

> **the weight of the beef in kilograms multiplied by 40.**

How many minutes are needed to cook a piece of beef weighing

a) 2 kilograms   **b)** 5 kilograms?

3 There is a rule for finding how far away thunderstorms are.

> **Count the seconds from the lightning to the thunder. Divide by 5. The answer is the distance in miles.**

    a) How far away is the storm if you count 5 seconds?

    b) How far away would it be if you counted 30 seconds?

4 As you go further up a mountain the atmosphere gets colder. There is a simple formula which tells you roughly how much the temperature will drop.

> **Temperature drop (°C) = height climbed in metres ÷ 200.**

If you climb up 800 m, about how much will the temperature drop?

5 To work out the number of rolls of wallpaper needed to paper a room some DIY books give the following rule.

> **Measure the distance round the edge of the room in feet.**
>
> **Call this number D.**
>
> **Measure the height of the room in feet.**
>
> **Call this number H.**
>
> **Multiply these two numbers together, i.e. find $D \times H$.**
>
> **Now divide by 50.**
>
> **The answer gives the number of rolls needed.**

    a) The distance round the room is 65 feet and the height is 10 feet. How many rolls are needed?

    b) The distance round the room is 50 feet and the height is 8 feet. How many rolls are needed?

Chapter 12 *Formulae*

Exercise 12.1A cont'd

6 This rule gives the approximate distance round a circle.

> **Multiply the radius by 6.**

   a) A circle has a radius of 6 cm. What is the distance round it?

   b) A round cake has a radius of 8 cm. What length of ribbon is needed to just go round it?

7 To find the number of words typed per minute, divide the number of words typed by the number of minutes taken. How many words per minute when:

   a) 120 words are typed in 4 minutes

   b) 220 words are typed in 5 minutes?

EXERCISE 12.1B

1 A cookery book gives this rule for cooking chicken.

Allow 15 minutes for each pound (lb) plus another 30 minutes.

   a) How long will a 3 lb chicken take to cook?

   b) Jill bought a 4 lb chicken. She put it in the oven at 4 p.m. When will the chicken be ready?

2 Mach number is a measure of speed.

An aeroplane travelling at Mach 1 is travelling at the speed of sound.

There is a rule for working out the Mach number.

Mach number = speed of sound in miles per hour divided by 760.

Calculate the Mach number of a plane travelling at 1520 mph.

3 To work out her weekly pay Jodi uses the formula:

Weekly pay = rate of pay per hour × hours worked + bonus.

Calculate Jodi's pay when she works for 30 hours at a rate of £6 per hour and earns a bonus of £30.

4 A rough rule to convert gallons into litres is:

Multiply the number of gallons by 9 and divide by 2.

   a) How many litres is 200 gallons?

   b) How many litres is 350 gallons?

5 To find the volume of a cone:

Multiply the area of the base by the height and then divide by 3.

Find the volume of a cone when the area of the base is 20 cm² and the height is 12 cm.

6 To change from English pounds to German Marks the rule is:

Multiply the number of pounds by 3.

How many German Marks would you get for £125?

Exercise 12.1B cont'd

7 To change from miles per hour to metres per second the rule is:

Multiply by 4 and divide by 9.

a) Change 45 metres per second into miles per hour.
b) Change 27 metres per second into miles per hour.

Key ideas

- A word formula shows how to work out a problem.
- Multiply numbers *before* you add or subtract so that $3 \times 4 + 5 = 17$ not 27.

Revision exercise

1 This angle measures 50°.

Estimate the size of the angles below. Write down your estimates and then check using a protractor/angle measurer.

a)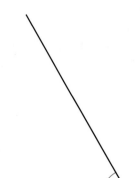

b)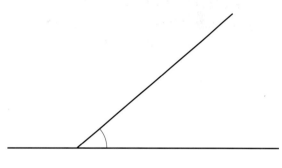

c)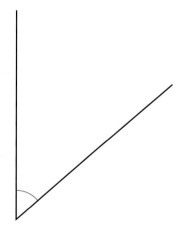

2

The leopard is 1·5 m tall. Estimate the height of the elephant and the giraffe.

3 This line 6 cm long.

───────────────────

Use this line to estimate the lengths of the following lines.

a) ─────────

b) ──────────────────────

c) ──────────────────────────────

4 Draw a full-size net for this cuboid.

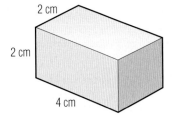

2 cm

2 cm

4 cm

5 Find the median and the mode of:

a) 4   3   15   9   7   6   11

b) 60 kg   12 kg   48 kg   36 kg   24 kg

6 A gardener measures the height of his sunflower plants as:

140 cm   123 cm   131 cm   89 cm
125 cm   123 cm   115 cm   138 cm

Find the median and the mode of these heights.

7 The cost of hiring a carpet cleaner is £15 deposit plus £4 per day. How much does it cost to hire the cleaner for four days?

8 A travel agent gives dollars for pounds (£) based on the formula:

£ $\xrightarrow{\times 1.5}$ $\xrightarrow{-10}$ dollars

Calculate how many dollars you get for **a)** £10   **b)** £40.

Drawing reflections

Emma is moving into a new bungalow.

The removal van is covering exactly half of the front of the bungalow. The hidden part is the same as the part you can see, only in reverse. What will the whole front of the bungalow look like?

There are two main ways to find out.

EXAMPLE 1

If you place a small mirror along the line of the back of the van you will see the other half of the bungalow in the reflection.

EXAMPLE 2

You can draw the reflection using tracing paper.

Place the tracing paper with one of its edges along the line of the back of the van.

Trace round the house.

Turn the tracing paper over, keeping the edges lined up, and this will complete the front of the bungalow.

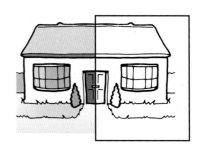

Reflection writing

EXERCISE 13.1A

Copy the following into your book and reflect in the dotted mirror line shown. Each one spells a word, gives a number or is a sum.

a)

RED

b)

HIDE

c)

ICE

d)

KICK

e)

CODE

f)

10

g)

22

h)

801

i)

10 + 12 = 21

j)

90 ÷ 10 = 9

EXERCISE 13.1B

Copy the following into your book and reflect in the dotted mirror line shown.
Each one spells a word, gives a number or is a sum.

a)

ᕼOᗺ

b)

ᐯIᗡ

c)

ᑢᕼOᐱᐸᕾ

d)

ᗡOᐯ

e)

ᕼOOᗡ

f)

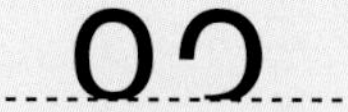

ᕦᕾ

g)

IOI

h)

ᕾᑎ

i)

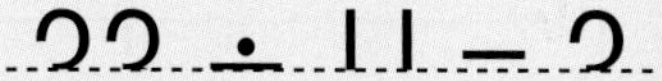

ᕾᕾ ÷ II = ᕾ

j)

O × ᕕ = ᕕ

k) See if you can find some more reflection words, numbers or sums.

Simple reflection

This section involves drawing reflections on squared paper.

EXAMPLE 3

Reflect the shape in the mirror line shown.

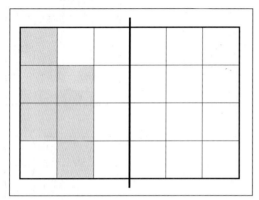 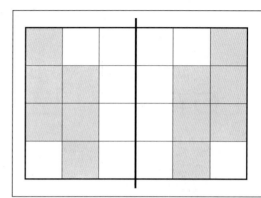

Notice that:

a) the shape is 'turned over' in a reflection

b) the 'image' is the same distance from the mirror line as the original shape but is on the opposite side of the mirror.

A small mirror or tracing paper can be used to help.

EXERCISE 13.2A

Copy these shapes onto squared paper.

Reflect each of the shapes in the mirror line shown.

a)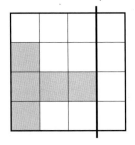

b)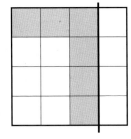

c)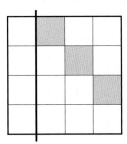

d)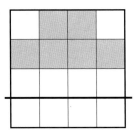

e)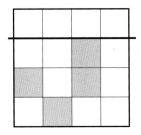

f)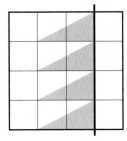

g)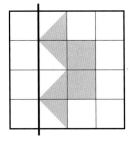

h)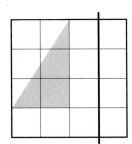

i)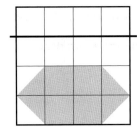

j)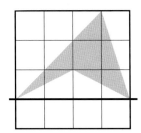

Chapter 13 *Drawing reflections*

EXERCISE 13.2B

Copy these shapes onto squared paper.

Reflect each of the shapes in the mirror line shown.

a)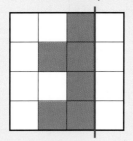

b)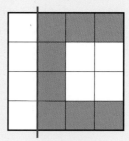

c)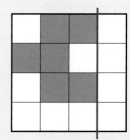

d)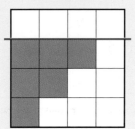

e)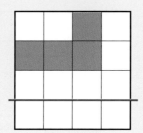

f)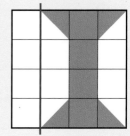

g)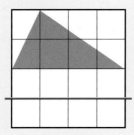

h)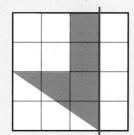

i)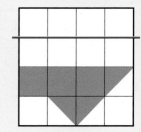

j)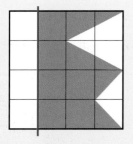

More difficult reflections

Type 1

Sometimes the original shape may have separate parts on both sides of the mirror line.

In these cases reflect each part separately onto the other side of the mirror line.

EXAMPLE 4

a) Reflect each part of the shape in the mirror line shown.

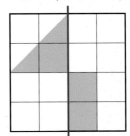

a)

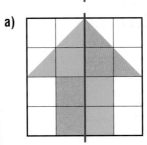

Type 2

You may be asked to reflect a shape in two mirror lines which cross at right-angles. If so:

a) reflect the given shape in one of the mirror lines, and then

b) reflect the result in the second mirror line.

EXAMPLE 5

a) Reflect this simple shape in both mirror lines shown.

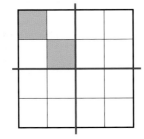

a) (i)

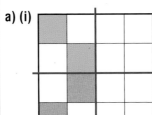

(ii)

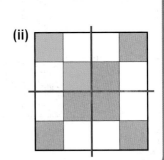

EXERCISE 13.3A

Copy the following shapes onto squared paper. Complete the reflections in the mirror lines shown.

a)

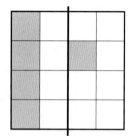

b)

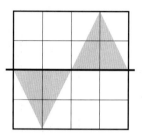

c)

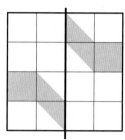

d)

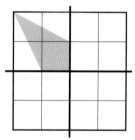

e)

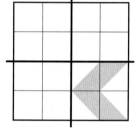

Chapter 13 *Drawing reflections*

EXERCISE 13.3B

Copy the following shapes onto squared paper. Complete the reflections in the mirror lines shown.

a)
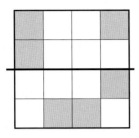

b)
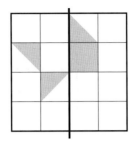

c)
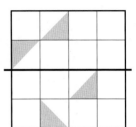

d)
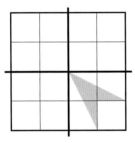

e)
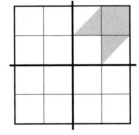

Planes of symmetry

Just as some two-dimensional shapes have lines of symmetry, so a three-dimensional shape can have one or more planes of symmetry.

EXAMPLE 3

This cuboid has three planes of symmetry. They are shown shaded.

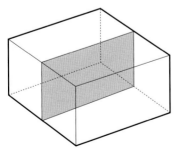
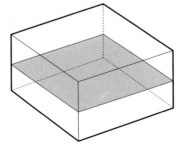
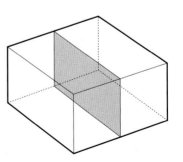

EXERCISE 13.4A

1   How many planes of symmetry does each of the following solids have?

a)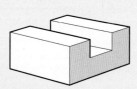

b)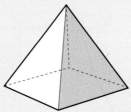

c)

2   How many planes of symmetry does each of the following solids have?

a) a cube

b) a circular

c) a cylinder

3   John has made a shape using seven cubes.

a) Use another seven cubes to make a shape which has a plane of symmetry.

b) Sketch your completed shape.

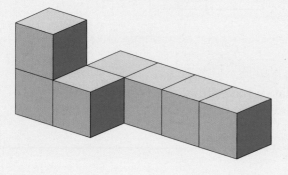

1 How many planes of symmetry does this shape have? (You might like to make the model to help you.)

2 This shape is made from 20 cubes. How many planes of symmetry does it have? (You might like to make the model to help you.)

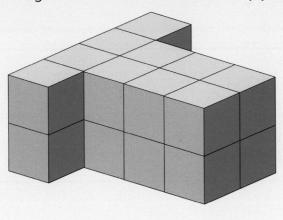

Exercise 13.4B cont'd

3 a) Here are three pictures of everyday objects.
How many planes of symmetry do they each have?

b) Find three more everyday objects that you think have planes of symmetry.
Sketch them and show their planes of symmetry.

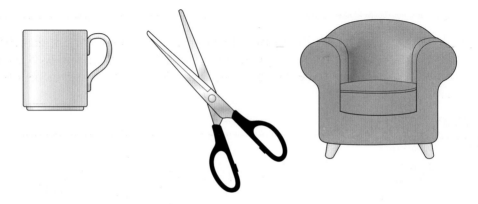

Key ideas

- 2D shapes may be reflected in a mirror line. The mirror line is a line of symmetry for the complete new shape.

- 3D shapes may have planes of symmetry.

14 Maps and plans

You should already know

● how to read scales and grids.

Maps and plans are often divided into **grids** and each grid is identified with a number to help people to find the position of objects. These numbers are called the **coordinates** or **grid references**. Maps usually show the direction of North, South, East and West.

EXAMPLE 1

Here is a map showing the position of some ships near a port.

a) Write down the grid references for all the ships.

b) From HMS *Vanguard* in which direction are
(i) the ferry?　**(ii)** the tanker?

a) The cruise ship's position is given the grid reference (20, 20). The first number tells you the position East or West, the second number the position North or South.
HMS *Vanguard's* position is (35, 30). It is 35 because it is halfway between the grid lines 30 and 40. The ferry is at (44, 15) and the tanker is at (22, 45)

b) From HMS *Vanguard*
(i) the ferry is South East
(ii) the tanker is North West.

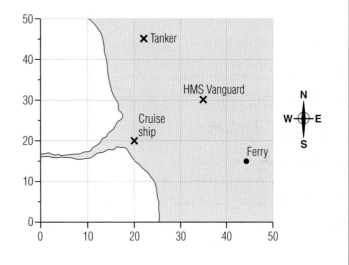

EXERCISE 14.1A

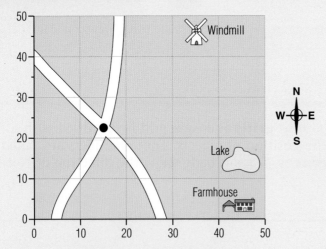

1 Write down the grid references of:

a) the roundabout   **b)** the windmill   **c)** the farmhouse.

d) The farm is due South of the lake. In which direction is the lake from the farm?

e) Write down the direction of the roundabout from the windmill.

Exercise 14.1A cont'd

2 This diagram shows an island.

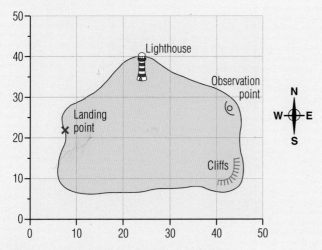

Write down the grid references for:

a) the landing **b)** the lighthouse **c)** the observation point

d) Between which two grid references are the cliffs?

Copy and complete this sentence, filling in the blanks.

The cliffs are .... of the lighthouse and the observation point is .... of the landing point.

3

This diagram shows a large lake.

a) What is due South of the island?

b) What is at (20, 0)?

c) Give the coordinates of boats B and C.

EXERCISE 14.1B

1 The map shows a park.
  a) What is at (50,10)?
  b) Write down the grid references for
  (i) tennis court
  (ii) seat
  c) Copy and complete
  The fountain is

  ....

  of the pond and

  ....

  of the greenhouse.

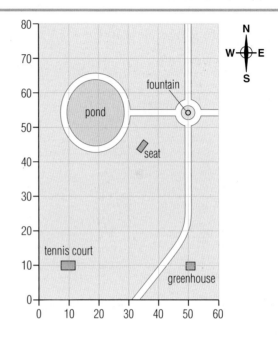

2

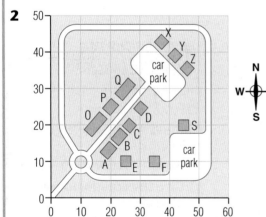

This is the plan of a shopping centre.
  a) What direction is F from the roundabout?
  b) Write down the grid references for
  (i) S   **(ii)** P.
  c) Which shop is at
  (i) (25,10)   **(ii)** (30, 25)?

Exercise 14.1B cont'd

3

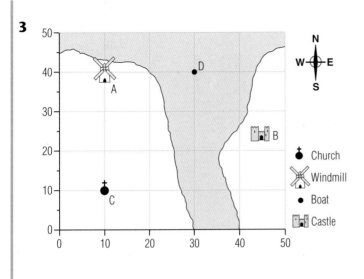

This diagram shows part of a coast where a river enters the sea.

a) What is due West of the boat at D?

b) Estimate the coordinates of the castle at B.

c) What is at the point (10,10)?

d) Give the coordinates of points A and D.

Key:
● Church
✳ Windmill
• Boat
🏰 Castle

Using maps

Are you lost? You need a map!

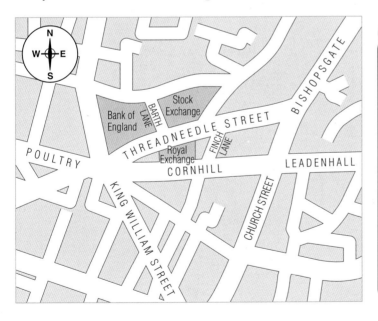

EXAMPLE 3

The map shows part of London.

Find the Bank of England.

Leave the Bank and turn LEFT on Threadneedle Street.

Pass a street on your left (Barth Lane).

What is the next building on your left? (Stock Exchange)

Take the next turn RIGHT. What street are you in? (Finch Lane)

Take the next RIGHT. In which direction are you walking? (West)

What building is on your right? (Royal Exchange)

EXERCISE 14.2A

1 This map shows part of Stratford-upon-Avon.

Copy and complete this journey.

Leave the Civic Hall and turn LEFT on Rother Road.

At crossroads turn RIGHT along ....

Take the next RIGHT into ...., walking in direction ....

At the next crossroads .... is on the corner.

2 Here is part of Harrogate.

Copy and complete this journey.

Leave Harrogate Station and walk North along ....

Turn .... into Cambridge Street.

Take the second on the RIGHT into .... and go into .... on your LEFT.

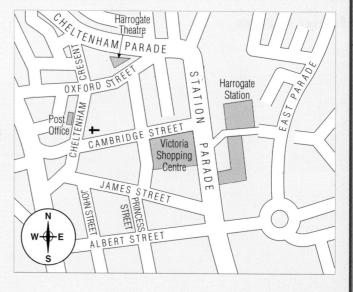

Exercise 14.2A cont'd

3 This is a map showing part
of St Andrews.

Write out a route from the
Cinema to the Town Hall.

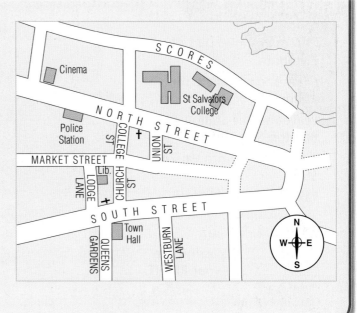

EXERCISE 14.2B

1 This map shows part of Plymouth.

Copy and complete this journey.

Leave the Bus Station on Brenton Side. Turn LEFT.

Turn .... into Vauxhall Street.

Take the next LEFT into ....

Go straight on into ...., passing the .... on your RIGHT.

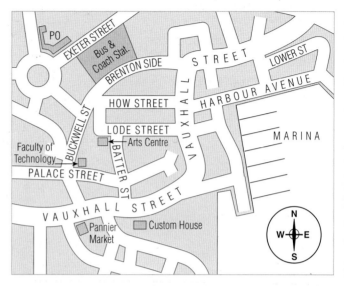

2 This is the centre of Huddersfield.

Write a route from the Station to the Kingsgate Shopping Centre.

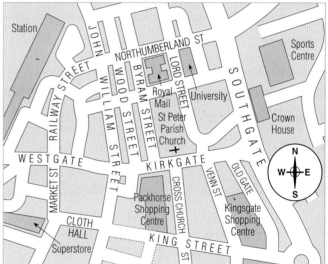

Exercise 14.2B cont'd

3 Here is part of Stoke-on-Trent.

Copy and complete this journey.

Leave the Car Park on Clough Street and turn RIGHT.

Take the first LEFT into ....

At the end of the street, turn LEFT into ....

You are now walking in direction ....

Take the third on the right into .... and pass .... on your LEFT.

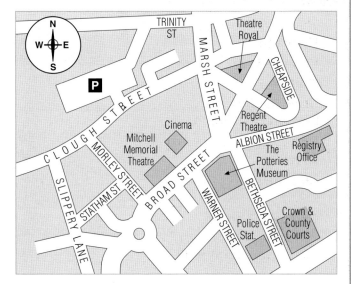

Key ideas

- Grid references are like coordinates with the East number first.
- On maps, directions can be given using compass points, or turning left or right.

Two-ways tables

You should already know

● how to add and subtract numbers.

Two-way tables are very useful for displaying complicated information. The best way to understand them is to work through this example.

EXAMPLE 1

80 girls in Year 11 have chosen whether they want to play hockey and/or netball during their PE lessons.

50 wanted to play hockey. 25 did not want to play netball. 35 wanted to play both hockey and netball.

This information can be put into a two-way table. Start by putting in the numbers you know.

	Hockey	Not hockey	Total
Netball	35		
Not netball			25
Total	50		80

Next, by addition and subtraction, you can work out what the missing numbers must be.

	Hockey	Not hockey	Total
Netball	35	this must be $55 - 35 = 20$	this must be $80 - 25 = 55$
Not netball	this must be $50-35 = 15$	this must be $25 - 15 = 10$	25
Total	50	this must be $20 + 10$ (or $80 - 50$) $= 30$	80

Now you know how many girls want, or don't want, to play each game.

EXERCISE 15.1A

1 Here is a two-way table showing the results of a car survey looking at makes and colours. Copy and complete the table.

	Vauxhall	Not Vauxhall	Total
Black	20		70
Not black		300	360
Total			

2 Mrs Fletcher asked the 270 pupils in Year 9 which theme park they wanted to go to for the end-of-year outing. The choices were Alton Towers or the American Adventure. 158 wanted to go to Alton Towers, 101 did not want to go to the American Adventure and 33 did not want to go to either place.

Complete this two-way table to show how many pupils were prepared to go to either theme park.

	Alton Towers	Not Alton Towers	Total
American Adventure			
Not American Adventure		33	101
Total	158		270

3 Here is a two-way table showing what language all the students opted to study in Year 10.

	French	Not French	Total
German	44	25	
Not German	50	20	
Total			

a) How many students were there in the year group?

b) How many wanted to study both French and German?

Exercise 15.1A cont'd

4   A cereal manufacturer tested what was thought to be an improved flavour breakfast cereal. Copy and complete the table showing the results of the test.

	Existing cereal	Improved cereal	Total
Liked	600	450	
Did not like	200	150	
Total			

a) How many people took part in the test?

b) How many people liked the new flavour?

c) Do you think the improved flavour cereal is better than the existing one? Give a reason for your answer.

5   At the world indoor athletics championships the USA, Germany and China won most medals.

For these three countries the medal table was:

	Gold	Silver	Bronze	Total
USA	31		10	
Germany	18	16		43
China		9	11	42
Total		43		

Copy and complete the table.

a) Which country won the most gold medals?

b) Which country won the most bronze medals?

EXERCISE 15.1B

1 Two tutor groups gave their choices for an activity day.

	Riding	Sport	Total
Male		18	
Female	15		
Total		25	48

Copy and complete the table.

2 Here is a two-way table showing which sports some students opted to play.

	Hockey	Not hockey	Total
Badminton	33		
Not badminton			39
Total	57		85

Copy and complete the table.

How many students did not want to play either hockey or badminton?

3 Here is a two-way table showing the results of a car survey.

Copy and complete the table.

	Japanese	Not Japanese	Total
Red	35	65	
Not red	72	438	
Total			

a) How many cars were surveyed?

b) How many Japanese cars were in the survey?

c) How many Japanese cars were a colour other than red?

d) What was the total number of red cars surveyed?

Exercise 15.1B cont'd

4 A drugs company has compared a new type of drug for hay fever with an existing drug.

Here is a two-way table showing the results of the trial.

	Existing drug	New drug	Total
Symptoms eased	700	550	
No change in symptoms	350	250	
Total			

Copy and complete the table.

a) How many people took part in the trial?

b) How many people using the new drug had their symptoms eased?

Key idea

● In two-way tables, adding the row totals should be the same as adding the column totals.

Revision exercise

1 Complete the reflection in the mirror line shown.

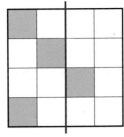

2 Copy these shapes and complete the reflections in the mirror lines shown.

a)

b)

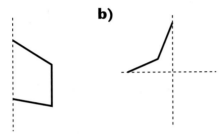

c)

3 How many planes of symmetry do the following objects have:

a) a cereal packet

b) a Toblerone box?

4 Look at the map below.

a) Write down the grid references for:
(i) Portrush
(ii) Cushendall
(iii) Cloughmills

b) (i) What town is almost due West of Dervock?

(ii) What tourist attraction is North East of Garvagh?

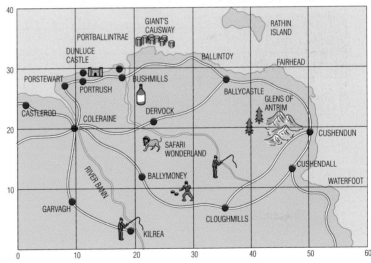

5 This table shows the results of a survey into people's preferences for flavours of crisp.

	Like salt and vinegar	Do not like salt and vinegar	Total
Like cheese and onion	130		180
Do not like cheese and onion		75	155
Total			

Copy and complete the table.

6 Form 7 is to have a day out at the end of the term. The pupils were asked to vote for their chosen place. The table shows some of their votes.

	Alton Towers	Legoland	Thorpe Park	Total
Girls	31	25		
Boys		11		75
Total	74		48	

Copy and complete the table.

Stage 1 – Page 1–134
Stage 2 – Page 135–242

A

acute angles — 152
addition — 10–13
    of decimals — 136
    of times — 186
    methods — 10, 11
algebra — 53–78
    defined — 53
    expressions — 53–55, 60
    function machines — 70–78
    letters, use of — 53–55, 60
    number patterns — 66–69
angles — 151–161
    acute — 152
    degrees — 151
    estimating — 154
    measuring — 151–154
    obtuse — 152
    protractors, use of — 153–154
    reflex — 152
    right — 152
area — 112, 116
    defined — 112
    estimating — 116
    irregular shapes — 116
    regular shapes — 112
average — 204–207
axes — 30

C

capacity (volume) — 120
chance (probability) — 18–21, 24
compass directions — 27, 227–230
compasses, shape construction — 93
coordinates — 30–38
    axes — 30
    graphical — 30
    maps — 227
    origin — 30
    plotting points — 30, 32
    points — 30
    x-axis — 30
    y-axis — 30
cube — 199
cuboid — 199, 203
    planes of symmetry — 224
cylinder — 199, 203

D

data representation — 124–130
    graphs — 128
    pictograms — 124
decagon — 92
decimal numbers — 137
    in probability — 175
decimals — 137, 140
    adding and subtracting — 138
    money, working with — 138
degrees — 151
digits — 1–5
    place value — 1–4
    whole numbers — 1
direction — 27–38
division — 15, 162–166
    large numbers — 164
    methods — 14
    remainder — 165
drawing angles — 151–161
    reflections — 215–226

E

enlargements — 95
    scale factor — 95
equilateral triangles — 90

INDEX

Stage 1 – Page 1–134
Stage 2 – Page 135–242

estimates
    angles   154
    lengths   191–195
even numbers   8
event   23, 24
expressions
    algebraic   60, 78
    lengths   53–55

F
figures, writing in   1
formulae   208–212
    in words   208–212
fractions   81–86
    as percentage   178
    finding   85
function machines   70–78

G
graphs, reading from   128
grid references   227
grids
    maps and plans   227
    multiplication and division   14, 15

H
hexagon   92

I
illustrating data
    graphs   30, 128–130
    pictograms   124–127
    pie charts   181–182

imperial units   101
isoceles triangles   90

K
kite   91

L
large numbers   4
lengths   101-105
    estimates   191–195
    measurement   101–103
letters, use in algebra   51–55, 60
listing   131–132

M
maps   227–235
    compass directions   227
    using   231–233
mass (weight)   169–171
    imperial units   169
    metric units   169
median   204–207
metric units   101–105
    mass   169
mirror lines   215–223
multiples of 5 and 10   9
multiplication   14, 162–166
    large numbers   177–182
    using grids   14

N
nets, 3-D shapes   199
number patterns   66–69
number facts   10–12

Stage 1 – Page 1–134
Stage 2 – Page 135–242

numbers	1–17
addition	10–13
and words	1, 4
digits	1–4
division	14–17, 85–86
facts	10–17
large	4
listing	131–132
multiplication	14–17
odd and even	8
place value	1, 4
rounding	6
subtraction	10–13
whole	1
numerator	81

O

obtuse angles	152, 154, 161
octagon	92
odd numbers	8
origin	30

P

parts, fractions as	81, 84–86, 87
patterns	
number	66–69
sequences	141
pentagon	92
percentage	178–184
as fractions	178
pie chart scales	179–180
perimeter	107–109
defined	107
formula	208
of composite shapes	109
of simple shapes	107
pictograms	124
pie charts	181–182
place value	1, 4, 137
table	4
planes of symmetry	223, 226
plotting points	30, 32
polygons	92
constructing	93–95
regular	92
position, sequences	145
probability	18–24, 172–177
definition	18, 172
scales	18–19
problem solving	79–87
reverse flow chart	301
trial and improvement	296
protractors	93, 153–156

Q

quadrilateral	91

R

reading graphs	128
rectangle	91
reflections	
drawing	215–226
difficult	221–223
mirror lines	215–223
planes of symmetry	223–226
simple	218–220
reflex angles	152
regular polygon	92
remainder	165
rhombus	91

Stage 1 – Page 1–134
Stage 2 – Page 135–242

right angles	152
right-angled triangle	90
rounding numbers	6
rules	
finding n	145
sequences	143

S

scale factor	95
scalene triangles	91
scales	39–50
pie charts	181–182
probability	18, 24
reading	44, 45
sequences	141–150
finding n	145
rules	143
term number	145
term value	145
solving problems	79–87
square	91
subtraction methods	10–12
symmetry, planes of	223–226

T

tables	
place value	3
two way	236–240
term number	145
term value	145
three dimensional shapes	199–200
net	199
planes of symmetry	223–226
time	185–189
addition	186

differences	187
elapsed	42, 43
finishing	43
telling	39–41
24-hour clock	185–187
trapezium	91
triangles	90–91
equilateral	91
isoceles	90
right-angled	90
scalene	91
24-hour clock	185–187
two-way tables	236–240

U

units 101–121	
imperial	101
metric	101–105

V

volume	119

W

weight. *See* mass	
whole numbers	1
words	
formulae in	208
telling time in	39

X

x-axis	30

Y

y-axis	30